Application of Antibody and Immunoassay for Food Safety

Application of Antibody and Immunoassay for Food Safety

Editors

Hongtao Lei
Zhanhui Wang
Sergei A. Eremin

MDPI • Basel • Beijing • Wuhan • Barcelona • Belgrade • Manchester • Tokyo • Cluj • Tianjin

Editors

Hongtao Lei
College of Food Science
South China Agricultural University
Guangzhou
China

Zhanhui Wang
Basic Veterinary Medicine Department, College of Veterinary Medicine
China Agriculture University
Beijing
China

Sergei A. Eremin
Department of Chemical Enzymology, Faculty of Chemistry
M.V.Lomonosov Moscow State University
Moscow
Russia

Editorial Office
MDPI
St. Alban-Anlage 66
4052 Basel, Switzerland

This is a reprint of articles from the Special Issue published online in the open access journal *Foods* (ISSN 2304-8158) (available at: www.mdpi.com/journal/foods/special_issues/antibody_immunoassay_food).

For citation purposes, cite each article independently as indicated on the article page online and as indicated below:

LastName, A.A.; LastName, B.B.; LastName, C.C. Article Title. *Journal Name* **Year**, *Volume Number*, Page Range.

ISBN 978-3-0365-3632-3 (Hbk)
ISBN 978-3-0365-3631-6 (PDF)

Contents

About the Editors

Hongtao Lei

Hongtao Lei is a full professor of South China Agricultural University (SCAU), Guangzhou, China. He obtained his bachelor degree in food science and engineering from Northwest Agriculture and Forestry University in 1997. Following this, he did a master research in food chemistry from September 1997 to June 2000 in SCAU, and then joined the College of Food Science, SCAU, as an assistant lecturer. From September 2003 to June 2006, he did PhD research in food safety in SCAU. From February 2009 to February 2010, he carried out one year of postdoctoral research at the Institute of Global Food Security, Queen's University of Belfast, U.K.. He was appointed as a professor in food science in December 2012, and as dean since June 2015. His research interest focuses on food quality and safety, molecular recognition, food authenticity, etc. He has coordinated more than 25 scientific research projects supported by national or local grant, and compiled 10 professional standards for product quality or analytical methods. He has published more than 100 peer reviewed articles in international journals in the food science area, such as Comprehensive Reviews in Food Science and Food Safety, Food Chemistry, Analytic Chemistry, Journal of Hazardous Materials, Biosensors and Bioelectronics, etc. His achievements were widely used in the field of rapid detection and were authorized more than 66 national invention patents. He has won six national and provincial awards due to his academic and teaching achievements. He is an associate editor of Chemical and Biological Technologies in Agriculture, editorial member of 10 journals such as Food and Agricultural Immunology, Chinese Food Science, etc. He is also the deputy director of National Teaching Steering Committee for Higher Education of Food Science and Engineering, Ministry of Education, deputy Chairman of the Agricultural Products Processing and Storage Engineering Committee, Chinese Society of Agricultural Engineering, etc.

Zhanhui Wang

Zhanhui Wang is a full professor of Veterinary Medicine at the China Agricultural University (Be, China). He completed his D.Phil. at the college of Veterinary Medicine. His interests focus on the following areas: the preparation and in vitro evolution of antibodies for small molecule such as veterinary drugs, mycotoxins and illegal additives; the development of antibody-based analytical techniques; and the antigen-antibody recognition mechanism. He has hosted or participated many multiple scientific research projects. He is the author of more than 120 publications in leading journals such as Anal Chem, Biosens Bioelectron, ACS Appl Mater Inter. He also developed more than 20 kinds of widely used rapid detection products and was authorized more than 50 national invention patents. He is a member of the editorial board of J Agric Food Chem, Food Agric Immunol, Sensors and Biomolecules. In addition, he served in several organizations including the director of BeKey Laboratory of Animal Derived Food Safety Detection Technology, deputy general secretary of Food Quality and Safety Testing Instruments and Technology Application branch of China Instrumentation Society, executive director of Veterinary Pharmacology and Toxicology branch of China Animal Husbandry and Veterinary Society.

Sergei A. Eremin

Prof. Sergei A. Eremin is head of Immunoassay group and leading researcher of the Department of Chemical Enzymology, Faculty of Chemistry, M.V.Lomonosov Moscow State University. He was educated in 1976 and obtained PhD in Organic Chemistry in 1982 and DSc degree in Biotechnology in 2005 from M.V.Lomonosov Moscow State University, Russia. His scientific interests lie in the field of immunoanalytical techniques for food safety and environmental control. He prepared immunoreagents and developed a fluorescence polarization immunoassay (FPIA) for detection of pesticides, mycotoxins, veterinary drugs, endocrine disruptors and other small molecule organic compounds. He is author of 279 papers in peer-reviewed journals and participated in several international projects covering different aspects of food safety and food-related risks. h-index = 34.

Preface to "Application of Antibody and Immunoassay for Food Safety"

Immunoassays are a class of analytical techniques wherein the reaction is based on highly specific molecular recognition between antibodies and antigens. Immunoassay has played a prominent role in the rapid detection of various analytes in food, including pesticides, veterinary drugs, heavy metals, hormones, allergens, food adulterants, natural components, biomarkers in food materials, etc.

A wide range of immunoassays have emerged endlessly, ranging from conventional enzyme-linked immunosorbent assays (ELISA) and point-of-care tests typically represented by lateral flow immunochromatography assay (LFIA) to biosensors with various principles, full integration of lab-on-a-chip platforms, microfluidics, sensibilization employing novel nanomaterials, miniaturization and interfacing of portable devices, especially emerging smart system technologies equipped with intelligent smart phones, etc. New technologies are promoting the development of immunoassays and their application in food safety.

Immunoreagents for food analysis are continuously developing during the last three decades, with hapten design leading to the possibility of using antibodies against non-common low molecular weight antigens and quantitative structure–activity relationship investigations, and structure biology approaches assisting toward a better understanding of the molecular recognition of epitope and antibody. In addition, novel antibodies such as various recombinant or fragment antibodies have contributed to the identification of various novel characteristics of antibodies in food safety.

We organized 11 articles to demonstrate the recent progress and broadening the novel knowledge about antibodies and immunoassays for the detection of chemical and biological analytes in food, and all of the 11 articles have already been published online in the Special Issue of {Foods}, "Application of Antibody and Immunoassay for Food Safety". Herein, we reprint this Special Issue as a book, to help readers to comprehensively and easily understand the current state concerning antibodies and immunoassays for food safty.

The book addressed several critical issues in application of antibodies and immunoassays for food safety. (1) Methodologies for the enhancement of sensitivity, rapidity, and reliability; (2) eliminating food matrix effect; (3) exploiting high-quality antibodies and new specific recognition elements, such as bispecific monoclonal antibody, broad-specific antibodies, and single chain antibody fragment (scFv) mutants; (4) preparing stable and strong signal labels to improve the accuracy and sensitivity of immunoassays; (5) multiple detection abilities as inevitable trends in the immunoassay development, etc.

Finally, we thank all the contributors for this book publishing, including all authors, academic and managing editors from the publisher MPDI.

Hongtao Lei, Zhanhui Wang, and Sergei A. Eremin
Editors

foods

MDPI

Editorial

Application of Antibody and Immunoassay for Food Safety

Hongtao Lei [1,*], Zhanhui Wang [2,3], Sergei A. Eremin [4] and Zhiwei Liu [1]

1 Guangdong Province Key Laboratory of Food Quality and Safety/National-Local Joint Engineering Research Center for Machining and Safety of Livestock and Poultry Products, College of Food Science, South China Agricultural University, Guangzhou 510642, China; liuzhiwei2021888@163.com
2 Beijing Advanced Innovation Center for Food Nutrition and Human Health, College of Veterinary Medicine, China Agricultural University, Beijing 100193, China; wangzhanhui@cau.edu.cn
3 Beijing Key Laboratory of Detection Technology for Animal-Derived Food Safety, Beijing Laboratory for Food Quality and Safety, Beijing 100193, China
4 Department of Chemical Enzymology, Faculty of Chemistry, M.V. Lomonosov Moscow State University, 119991 Moscow, Russia; eremin_sergei@hotmail.com
* Correspondence: hongtao@scau.edu.cn

This Special Issue of Foods, Application of Antibody and Immunoassay for Food Safety, contains ten papers that were refereed and selected in accordance with the usual editorial standards of the journal.

The aim of this Special Issue is to advance the current state of knowledge concerning antibodies and immunoassays for the detection of chemical and biological analytes such as food contaminants, food fraud, and so on, in the field of food safety.

Food safety is of critical societal importance for producers, food agencies, regulatory bodies and consumers. Therefore, there is a need to develop fast, sensitive, reliable, cost-effective, and easy-to-use analytical techniques for the protection of food safety and quality. Currently, instrumental analysis methods are commonly used for food safety purposes, such as high-performance liquid chromatography (HPLC), high-performance liquid chromatography–tandem mass spectrometry (HPLC-MS/MS), etc. Although the aforementioned assays are validated, sensitive, and reliable, they are unsuitable for application in rapid screening and field detection owing to the requirements of expensive apparatus, time-consuming operation, and highly skilled personnel. Immunoassays, a class of analytical techniques based on the specific recognition between antibody and antigen, are preferable to overcome these obstacles because of their high sensitivity, specificity, rapidity and cost-effectiveness, which allow them to play a prominent role in the rapid detection of various analytes in food safety [1].

To pursue higher sensitivity, advances have been made to improve the analytical sensitivity of immunoassays. In particular, chemiluminescent immunoassay (CLEIA) and fluoroimmunoassay (FIA) are two commonly proposed methods to meet the needs of strict screening. Ou et al. [2] prepared a monoclonal antibody against aristolochic acid I (AA-I) and applied it in CLEIA and FIA for the highly sensitive determination of aristolochic acid I (AA-I) in foods (slimming capsule, slimming tea, and pleurotus ostreatus). The proposed CLEIA showed higher sensitivity compared with conventional ELISA. On the other hand, a novel fluorescent probe, carbon dots, was synthesized and employed in FIA, which exhibited a five-fold greater enhancement in sensitivity than CLEIA. Moreover, the accuracy and practicability of CLEIA and FIA were verified by the standard instrument method, indicating that both were sensitive, rapid, and easy to use, making them effective tools for screening AA-I in related products. Additionally, there are also various emergent strategies that address the poor sensitivity of immunoassays, including novel signal labels (i.e., nanozymes and magnetic-loaded nanoparticles), unique antibody with unique nature, and heterologous strategies adjusting the binding capability of the competitive antigen, as well as in combination with innovative detection platform (i.e., microfluidic detection platform, smart detection systems, and a detection platform combined with molecular biology).

Citation: Lei, H.; Wang, Z.; Eremin, S.A.; Liu, Z. Application of Antibody and Immunoassay for Food Safety. *Foods* **2022**, *11*, 826. https://doi.org/10.3390/foods11060826

Received: 28 February 2022
Accepted: 7 March 2022
Published: 14 March 2022

Despite enormous sensitivity-enhanced strategies for immunoassays, the unwanted interference from food matrix is still a major factor affecting assay sensitivity due to the complexity and variability of matrix compounds in food samples, which might greatly affect the immunological reaction. Assessing the matrix interference on the assay sensitivity is thus an important issue in the development of methods. Burkin et al. [3] firstly evaluated the influence of avidin (AVI) and biotin (B7) contained in food matrices on two kinds of (Strept)avidin–biotin-based enzyme-linked immunosorbent assays (ELISAs) for bacitracin (BT) and colistin (COL) determination, with simultaneous assessment of extraneous AVI/B7 and AVI/B7 from different matrices (egg, infant milk formulas enriched with B7, and chicken and beef liver). Summarizing the experience of the present study, it is recommended to avoid immunoassays based on avidin–biotin interactions when analyzing samples containing these endogenous factors or enriched with B7.

Immunoassays, especially for ELISA, are generally heterogeneous, and involve repeated washing and a certain degree of reaction time. In contrast, fluorescence polarization immunoassay (FPIA) is a homogeneous assay format without separation or washing, giving the advantages of rapidity, reliability, and ease of use. It is based on the competition between an analyte and a fluorescein-labeled tracer for binding antibody. Zhang et al. [4] established an FPIA for 4,4′-dinitrocarbanilide (DNC) in chicken samples, with favorable sensitivity, specificity, cost, time, and reliability. The sensitivity of the developed FPIA was significantly improved by optimizing the selection of 25 tracers, tracer–antibody pairs, and physical and chemical reaction conditions. Furthermore, the reliability and robustness of the assay were successfully demonstrated for the analysis of DNC in chicken muscle matrices. The total analysis time, including sample pretreatment, was less than 40 min, which has not yet been achieved in other immunoassays for DNC. Up to now, many FPIA for other analytes such as mycotoxins, pesticides, antibiotics, and so on, have been tested and compared favorably with instrumental reference methods.

Compared with the ELISA-based and FPIA assays mentioned above, another immunoassay, namely lateral flow immunochromatography assay (LFIA), has gained increasing popularity because of its simple operation, rapidity, sensitivity, and cost-effectiveness. Li et al. [5] focus on the development of a rapid, convenient and sensitive LFIA based on traditional Au nanoparticles (AuNPs) for furosemide in slimming health foods, and the results could be read by the naked eye within 12 min (including sample pretreatment). The qualitative limit of detection (LOD) of the AuNPs-LFIA was 1.0–1.2 μg/g in slimming health foods. The developed method showed high consistency with liquid chromatography–tandem mass spectrometry (LC-MS/MS), and no false positive or false negative results were found in spiked slimming health foods. However, AuNPs-LFIA is known to have limited sensitivity because of AuNPs' narrow particle size range and poor colloid stability. Wu et al. [6] described a background fluorescence-quenching immunochromatographic assay (bFQICA) in which AuNPs were used to quench the fluorescence of a background fluorescence baseboard instead of using the colorimetric method. Such a method was optimized, validated, and applied in the rapid on-site detection of nitrofurazone metabolite of semicarbazide (SEM) residues in animal-derived foods (egg, chicken, fish, and shrimp). Indeed, compared with the traditional AuNP-LFIA method, the detectability of the bFQICA method was higher, and the detection time was shortened compared with heterogeneous reactions such as ELISA. In addition, the quantitative results of SEM can be directly displayed by using a portable fluorescence immunoquantitative analyzer and a QR code with a built-in standard curve, which is efficient and convenient. Additionally, the signal label is a vital factor in the performance of LFIA. Novel nanoparticle labels have been introduced to obtain satisfactory sensitivity. Chen et al. [7] designed a chiral carbon containing a structure similar to that of propiconazole, and a polyclonal antibody that specifically recognizes propiconazole was obtained for the first time. Based on this antibody, a time-resolved fluorescence microspheres lateral flow immunochromatographic assay (TRFMs-LFIA) was developed, optimized, and evaluated for its sensitivity, specificity, and recovery. The analysis of blind real-life samples (brassica campestris, lettuce, and romaine lettuce) showed a good agreement with results obtained using HPLC-MS/MS. Of course, there are many other nanoparticle labels have been synthesized for the enhancement of LFIA performance. Despite these advances, some still need to be improved in order to enhance their high-throughput capacity in a single assay, and to move toward miniaturization involves the use of mobile devices such as smartphones.

Some scholars have also focused on preparing specific recognition molecules with unique performance, such as broad specificity, low molecule weight or small size (single-chain variable fragment (ScFv)), disulfide-stabilized antibodies, antigen-binding fragment (Fab), nanobodies, bispecific monoclonal antibodies (BsMAbs), aptamers, molecularly imprinted polymers, etc. In recent years, BsAbs, broad-spectrum antibodies, and ScFv have been increasingly favored by researchers in the field of immunoassays.

Compared with a single-specific antibody which can only recognize one antigen in a complex food matrix, a bispecific monoclonal antibody (BsMAb) with two distinct antigen-binding sites could recognize two different target analyses, which is more efficient, convenient, and economical. Lu et al. [8] successfully prepared BsMAb against aflatoxin B1 (AFB1) and ochratoxin A (OTA), and developed a novel and efficient immunoaffinity column (IAC) based on BsMAb for the rapid and effective extraction of AFB1 and OTA with a one-time extraction from corn and wheat samples. Then, the ELISA for AFB1 and OTA were applied, combined with IAC, with a satisfactory matrix effect elimination effect and recovery rate. The development of BsMAb has opened a whole new field in multi-analyte detection. Future advances will include, but not be limited to, exploiting new methods based on BsMAb, and novel techniques of antibody development that will allow for two or more targets, as well as cheaper and faster analysis methods.

An immunoassay based on a broadly specific antibody is an emerging trend in the sensitive and simple detection of a group of similar compounds in a single assay. Shao et al. [9] designed and synthesized an unreported hapten, 5-(propylthio)-1H-benzo[d]imidazol-2-amine, which maximally exposed the characteristic sulfanyl group of albendazole (ABZ) to the animal immune system to induce the expected antibody. One mAb that can simultaneously detect the sum of ABZs (ABZ and its metabolites, i.e., $ABZSO_2$, ABZSO, and $ABZNH_2SO_2$) was obtained. The results of computational chemistry methods revealed that the hydrophobicity and conformation of a characteristic group of molecules might be the key factors that together influence the antibody recognition of these analytes. Furthermore, the practicability of the developed ELISA was verified by detecting ABZs in spiked milk, beef, and liver samples with recoveries. Zhang et al. [10] explored and developed novel haptens using molecular modeling to prepare broad-spectrum mAbs against brevetoxin 2 (PbTx-2), 1 (PbTx-1), and 3 (PbTx-3), followed by an ELISA method to detect brevetoxins in oyster samples was developed. In particular, the differences between the haptens of PbTx-2-CMO and PbTx-2-HS were evaluated using molecule alignment and electrostatic potential analysis. The results highlight that the spacer HS arm of PbTx-2-HS formed a specific spatial conformation with the parent nucleus structure, non-conducive to the production of high-affinity antibodies against the target, while PbTx-2-CMO was the ideal hapten to be used for antibody production due to its similar structure to the target, which was also further verified by antibody production and characterization. Besides ideal specificity and recovery rate, the sensitivity of the proposed ELISA based on such a mAb was higher than that of the high-resolution LC-MS, providing a useful method for monitoring PbTxs in oyster samples. The two studies above revealed that hapten design is an important feature when preparing antibodies against multiple target compounds. Importantly, molecular modeling and theoretical tools may assist immunochemists to find the most appropriate hapten chemical structure for broad-spectrum antibody production.

Single-chain variable fragments (scFv), as one of the most common formats of recombinant antibody, possess only one chain of the complete antibody while maintaining antigen-specific binding abilities, and can be expressed in a prokaryotic system. Moreover, it can be easily engineered with enhanced affinity and selectivity. Wang et al. [11] constructed an immunized mouse phage display single-chain variable fragment (scFv) library for the screening of recombinant anti-ciprofloxacin single-chain antibody for the detection of ciprofloxacin (CIP) in animal-derived food. The highest positive scFv-22 was expressed in *E. coli* BL21. Specifically, its recognition mechanisms were studied using the molecular docking method, and directional mutagenesis was performed for sensitivity improvement. The results of the established icELISA demonstrate that the ScFv mutant showed 16.6-fold improved sensitivity compared with parental scFv. Although scFvs have already found widespread use in clinical therapy and imaging procedures in the past decades, the use of such antibody fragments can provide clear benefits in terms of assay performance and relatively easy preparation in comparison to monoclonal and polyclonal antibodies, which will most probably lead to the increased use of recombinant antibodies in analytical applications in the near future.

To conclude, the present Special Issue addresses several critical issues, ranging from methodologies for performance enhancement (sensitivity, rapidity, and reliability), the assessment of food matrixes, and the development of specific recognition elements (BsMAb, broad-spectrum antibodies, and ScFv mutants). To satisfy higher detection requirements, the development of ultrasensitive and accurate immunoassays with multiple detection abilities is a growing trend. Exploiting high-quality antibodies and new specific recognition elements is particularly important to assay performance. Additionally, preparing stable and strong signal labels is necessary for improving the accuracy and sensitivity of immunoassays. Moreover, multiplex testing technologies are also inevitable trends in the development of immunoassays.

Finally, we thank the authors for their valuable contributions to this Special Issue.

Author Contributions: Conceptualization, H.L., Z.W., S.A.E.; Investigation, analysis and revision, Z.W., S.A.E.; Writing—original draft, Z.L., H.L. Resource, H.L. All authors have read and agreed to the published version of the manuscript.

Funding: This work was financially supported by the National Key Research and Development Program of Thirteenth Five-Year Plan (No. 2017YFC1601700), the National Scientific Foundation of China (31871883, U1301214), HeYuan Planned Program in Science and Technology (210115091474673), Generic Technique Innovation Team Construction of Modern Agriculture of Guangdong Province (2021KJ130), Lingnan Modern Agricultural Science and Technology experiment project of Guangdong Provinc (LNSYSZX001).

Conflicts of Interest: The authors declare no conflict of interest.

References

1. Zhang, H.; Li, B.; Liu, Y.; Chuan, H.; Liu, Y.; Xie, P. Immunoassay technology: Research progress in microcystin-LR detection in water samples. *J. Hazard. Mater.* **2022**, *424*, 127406. [CrossRef] [PubMed]
2. Ou, A.-F.; Chen, Z.-J.; Zhang, Y.-F.; He, Q.-Y.; Xu, Z.-L.; Zhao, S.-Q. Preparation of Anti-Aristolochic Acid I Monoclonal Antibody and Development of Chemiluminescent Immunoassay and Carbon Dot-Based Fluoroimmunoassay for Sensitive Detection of Aristolochic Acid I. *Foods* **2021**, *10*, 2647. [CrossRef] [PubMed]
3. Burkin, M.A.; Galvidis, I.A.; Eremin, S.A. Influence of Endogenous Factors of Food Matrices on Avidin–Biotin Immunoassays for the Detection of Bacitracin and Colistin in Food. *Foods* **2022**, *11*, 219. [CrossRef] [PubMed]
4. Zhang, Q.; Zou, M.; Wang, W.; Li, J.; Liang, X. Design, Synthesis, and Characterization of Tracers and Development of a Fluorescence Polarization Immunoassay for Rapid Screening of 4, 4′-Dinitrocarbanilide in Chicken Muscle. *Foods* **2021**, *10*, 1822. [CrossRef] [PubMed]
5. Li, Y.Y.; Xie, H.H.; Wang, J.; Li, X.M.; Xiao, Z.L.; Xu, Z.L.; Lei, H.T.; Shen, X. Lateral Flow Immunochromatography Assay for Detection of Furosemide in Slimming Health Foods. *Foods* **2021**, *10*, 2041. [CrossRef] [PubMed]
6. Wu, Y.P.; Wang, J.; Zhou, Y.; Qi, Y.H.; Ma, L.C.; Wang, X.N.A.; Tao, X.Q. Quantitative Determination of Nitrofurazone Metabolites in Animal-Derived Foods Based on a Background Fluorescence Quenching Immunochromatographic Assay. *Foods* **2021**, *10*, 1668. [CrossRef] [PubMed]
7. Chen, B.; Shen, X.; Li, Z.; Wang, J.; Li, X.; Xu, Z.; Shen, Y.; Lei, Y.; Huang, X.; Wang, X. Antibody Generation and Rapid Immunochromatography Using Time-Resolved Fluorescence Microspheres for Propiconazole: Fungicide Abused as Growth Regulator in Vegetable. *Foods* **2022**, *11*, 324. [CrossRef] [PubMed]
8. Lu, D.; Wang, X.; Su, R.; Cheng, Y.; Wang, H.; Luo, L.; Xiao, Z. Preparation of an Immunoaffinity Column Based on Bispecific Monoclonal Antibody for Aflatoxin B1 and Ochratoxin A Detection Combined with ic-ELISA. *Foods* **2022**, *11*, 335. [CrossRef] [PubMed]

9. Shao, S.; Zhou, X.; Dou, L.; Bai, Y.; Mi, J.; Yu, W.; Zhang, S.; Wang, Z.; Wen, K. Hapten Synthesis and Monoclonal Antibody Preparation for Simultaneous Detection of Albendazole and Its Metabolites in Animal-Origin Food. *Foods* **2021**, *10*, 3106. [CrossRef] [PubMed]
10. Zhang, X.; Ding, M.; Zhang, C.; Mao, Y.; Wang, Y.; Li, P.; Jiang, H.; Wang, Z.; Yu, X. Development of a New Monoclonal Antibody against Brevetoxins in Oyster Samples Based on the Indirect Competitive Enzyme-Linked Immunosorbent Assay. *Foods* **2021**, *10*, 2398. [CrossRef] [PubMed]
11. Wang, F.; Li, N.; Zhang, Y.; Sun, X.; Hu, M.; Zhao, Y.; Fan, J. Preparation and Directed Evolution of Anti-Ciprofloxacin ScFv for Immunoassay in Animal-Derived Food. *Foods* **2021**, *10*, 1933. [CrossRef] [PubMed]

foods

MDPI

Communication

Preparation of Anti-Aristolochic Acid I Monoclonal Antibody and Development of Chemiluminescent Immunoassay and Carbon Dot-Based Fluoroimmunoassay for Sensitive Detection of Aristolochic Acid I

Ai-Fen Ou [1,2], Zi-Jian Chen [3], Yi-Feng Zhang [3], Qi-Yi He [1], Zhen-Lin Xu [3] and Su-Qing Zhao [1,*]

1 Department of Pharmaceutical Engineering, School of Biomedical and Pharmaceutical Sciences, Guangdong University of Technology, Guangzhou 510006, China; ouaifen@gcp.edu.cn (A.-F.O.); chesto36@163.com (Q.-Y.H.)
2 Department of Food, Guangzhou City Polytechnic, Guangzhou 510006, China
3 Guangdong Provincial Key Laboratory of Food Quality and Safety/Guangdong Laboratory of Lingnan Modern Agriculture, South China Agricultural University, Guangzhou 510642, China; guangdongchenzj@163.com (Z.-J.C.); zhyifengscau@163.com (Y.-F.Z.); jallent@163.com (Z.-L.X.)
* Correspondence: sqzhao@gdut.edu.cn

Abstract: Aristolochic acid (AA) toxicity has been shown in humans regarding carcinogenesis, nephrotoxicity, and mutagenicity. Monitoring the AA content in drug homologous and healthy foods is necessary for the health of humans. In this study, a monoclonal antibody (mAb) with high sensitivity for aristolochic acid I (AA-I) was prepared. Based on the obtained mAb, a chemiluminescent immunoassay (CLEIA) against AA-I was developed, which showed the 50% decrease in the RLU_{max} (IC_{50}) value of 1.8 ng/mL and the limit of detection (LOD) of 0.4 ng/mL. Carbon dots with red emission at 620 nm, namely rCDs, were synthesized and employed in conventional indirect competitive enzyme-linked immunosorbent assay (icELISA) to improve the assay sensitivity of a fluoroimmunoassay (FIA). Oxidized 3,3′′,5,5′′-tetramethylbenzidine dihydrochloride (oxTMB) can quench the emission of the rCDs through the inner-filter effect; therefore, the fluorescence intensity of rCDs can be regulated by the concentration of mAb. As a result, the assay sensitivity of FIA was improved by five-fold compared to CLEIA. A good relationship between the results of the proposed assays and the standard ultra-high performance liquid chromatography-triple quadrupole mass spectrometer (UPLC-QQQ-MS/MS) of real samples indicated good accuracy and practicability of CLEIA and FIA.

Keywords: aristolochic acid I; monoclonal antibody; computer-assisted simulation; chemiluminescent immunoassay; fluoroimmunoassay

Citation: Ou, A.-F.; Chen, Z.-J.; Zhang, Y.-F.; He, Q.-Y.; Xu, Z.-L.; Zhao, S.-Q. Preparation of Anti-Aristolochic Acid I Monoclonal Antibody and Development of Chemiluminescent Immunoassay and Carbon Dot-Based Fluoroimmunoassay for Sensitive Detection of Aristolochic Acid I. *Foods* **2021**, *10*, 2647. https://doi.org/10.3390/foods10112647

Academic Editor: Thierry Noguer

Received: 27 September 2021
Accepted: 28 October 2021
Published: 1 November 2021

1. Introduction

Aristolochic acids (AAs) are a mixture of structurally related nitrophenanthrene carboxylic acids, mainly consisting of aristolochic acid I (AA-I) and aristolochic acid II (AA-II), which exist in Aristolochia spp. [1,2], a kind of Chinese herb. Moreover, these herbs can be used as raw materials of some drug homologous and healthy foods, and even dietary supplements [3–5]. However, it has been reported that AA showed toxicity to humans owing to carcinogenesis [2,6–8], nephrotoxicity [9–11], and mutagenicity [9,12]. Some cases reported that the intake of slimming products containing AA resulted in nephropathy [13,14]. Many countries have prohibited products containing AAs. Therefore, it is necessary to develop effective methods for detecting and monitoring AA in related food products.

For the analysis of AA, the main detection method is the conventional instrumental method, high performance liquid chromatography (HPLC) [15–17], due to the high accuracy and high reproducibility. Nevertheless, it is a challenge that the instrumental method is limited by high cost, the need for professional operators, and a long turnaround time. Therefore, immunoassay was proposed in this study because of the advantages of rapidness, high-throughput, sensitivity, low-cost, simple pretreatment requirement. It is easy-to-use and has been widely applied in fields of food analysis [18–24].

The most common and mature detection technology for the immunoassay of AAs is conventional indirect competitive enzyme-linked immunosorbent assay (icELISA); however, the sensitivity of icELISA cannot meet the needs of strict screening. In this study, a chemiluminescent immunoassay (CLEIA) was developed concerning its higher sensitivity compared with conventional icELISAs [25,26]. On the other hand, a fluoroimmunoassay (FIA) is a potential methodology through its advantages, including high sensitivity, real-time, fast response, and low cost to improve the method sensitivity [27]. As a novel fluorescent probe, the carbon dots (CDs) exhibit superiority of remarkable fluorescent properties, simple preparation, good biocompatibility, and easy functionalization [28], which can be employed for FIA. For FIA development, most previous publications reported a phosphate-triggered method to recover the fluorescence of CDs [29–31]. Nevertheless, alkaline phosphatase (ALP) activity requires more than 30 min for the catalyst, which is not satisfactory. For overcoming this time-consuming step, we employed horseradish peroxidase (HRP) and developed red CDs (rCDs)-based FIA. Compared with ALP, the activity of HRP is higher and its catalysate-oxidized 3,3′′,5,5′′-tetramethylbenzidine dihydrochloride (oxTMB) can quench rCDs. Based on the above principle, we developed HRP and rCDs based on FIA for AAs analysis.

2. Materials and Methods

2.1. Reagents and Animals

Standards of AA-I, AA-II, AA-III, AA-IV, and its analogs were purchased from Yuanye Co. Ltd. (Shanghai, China). Citric acid, urea, N,N-dimethylformamide (DMF), 1-(3-dimethylaminopropyl)-3-ethylcarbodiimide hydrochloride (EDC), N-hydroxysuccinimide (NHS) were supplied by Heowns Chemical Technology Co., Ltd. (Tianjin, China). The ovalbumin (OVA), keyhole limpet haemocyanin (KLH), and bovine serum albumin (BSA) were supplied by Sigma (Shanghai, China). The incomplete and complete Freund's adjuvants were purchased from Merck Co. Ltd. (Shanghai, China). The TMB and chemiluminescent substrate solution were obtained from Yuanye Co. Ltd. (Shanghai, China). Protein G resin and a secondary antibody (goat anti-mouse IgG, HRP conjugated) were obtained from TransGen Biotech Co. Ltd. (Beijing, China).

Bal b/c female mice were purchased from the Guangdong Medical Experimental Animal Centre and raised at the Animal Experiment Centre of South China Agriculture University (Animal Experiment Ethical Approval Number: 2019054, Figure S1).

2.2. Instruments

Multiskan FC microplate reader (Thermo Fisher, Shanghai, China) was used to measure absorbance values. The fluorescence was measured at emission (Em) wavelengths of 620 nm with excitation (Ex) wavelength of 540 nm using a SpectraMax i3 microplate reader (Molecular Devices, USA). A NanoDrop2000c spectrophotometer (Thermo Fisher, Shanghai, China) was used for concentration measurement and UV spectrum characterization.

2.3. Production of Monoclonal Antibody

Since AA-I is the main compound of AAs, it was directly conjugated to a carrier protein to prepare immunogen and coating antigen. The conjugation procedure was referred to in our previous study [32], and the details are summarized in Supporting Information. The artificial antigens were characterized by ultraviolet visible (UV-vis) spectral. The molar ratio between hapten and carrier protein was obtained by MALDI-TOF-MS.

The produced artificial antigens were used for animal immunization described in our previous study [32]. The production of mAb was followed by our previous publication [33]. The obtained mAb was purified by protein G and stored at −20 °C.

2.4. Development of Chemiluminescent Immunoassay

A serial concentration of AA-I (50 μL) and 50 μL of mAb solution was added to each well for 40 min incubation at 37 °C. Afterward, HRP-conjugated secondary antibody was added (100 μL/well) after five times washing with PBST (PBS with 0.5 ‰ Tween-20) for 30 min incubation at 37 °C. Then the chemiluminescent substrate solution was added (100 μL/well), and the RLU value was measured after a 1 min reaction. The calibration curve was fitted by sigmoidal fitting using the percent binding of mAb in the wells (RLU/RLU_0) against the logarithm of the AA-I concentration. The 50% decrease in the RLU_{max} (IC_{50}) value was calculated using Origin 8.5. The optimal conditions were confirmed from the IC_{50} value, including the optimal pH, coating antigen/antibody, phosphate, and Tween-20 concentration.

2.5. Development of Fluoroimmunoassay

The synthesis of rCDs is summarized in Supporting Information for the development of FIA. It was the same as CLEIA except using TMB as substrate and polystyrene transparent microplate. After oxTMB generation, the solution of each well was mixed with 50 μL of NaOH (1 mM, pH 11) to adjust the pH. Then, rCDs (50 μL) were added and mixed quickly. One hundred microliters of the mixture were transferred to black polystyrene opaque microplate, and the fluorescence signal was measured with Ex 540 nm and Em 620 nm.

2.6. Recovery Test

Samples (drug homologous and foods) were obtained from a supermarket in Guangzhou city. Samples (600, 300, and 150 ng/g) were ground into a powder using a stainless-steel grinder. The AA-I was added to samples (5.0 g) and mixed with methanol (5 mL) for 30 min ultrasonic water bath treatment. Afterward, the mixture was centrifuged at 4000 rpm (2057× *g*) for 10 min. The extraction solution was dried by nitrogen flow and redissolved in an equivalent volume of 0.01 M pH 5.4 PBS (PB with 75 mM NaCl) and employed for CLEIA and FIA. For UPLC-QQQ-MS/MS, the redissolved solutions were filtered by 0.22 μm cellulose membrane before analysis. The details of UPLC-QQQ-MS/MS are summarized in Table S1.

3. Results

3.1. Characterization of Antisera

In this study, AA-I was conjugated to a carrier protein directly. The ultraviolet scanning showed that the synthesized artificial antigens exhibited the characteristic absorption peak of AA-I and carrier proteins, suggesting the successful preparation of artificial antigens (Figure S2). The prepared immunogen and coating antigen were further utilized to prepare anti-AA-I mAb. The strategy is shown in Figure 1. The mouse antisera characterization is summarized in Table 1. The highest titer and sensitivity were observed for mouse 4 using AA-I-KLH, which was chosen for the production of mAb. After subclone and hybridoma screening, the most sensitive cell lines 3A3 were obtained and used for ascites preparation (Table S2). Finally, the mAb was obtained after ascites purification and was stored at −20 °C after concentration measurement by Nanodrop.

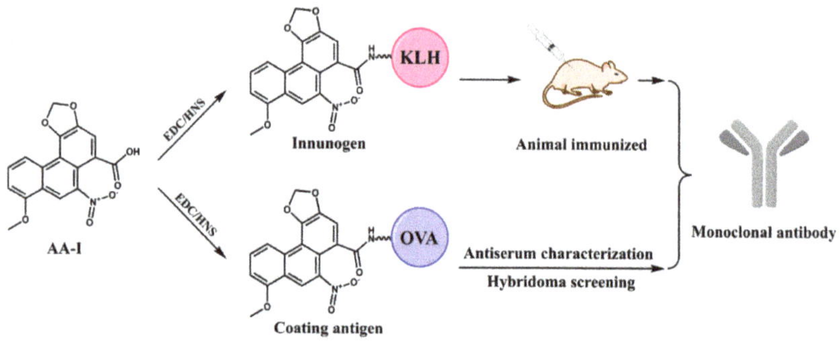

Figure 1. The strategic schema of mAb production. AA-I: aristolochic acid I; EDC: 1-(3-dimethylaminopropyl)-3-ethylcarbodiimide hydrochloride; NHS: N-hydroxysuccinimide; KLH: keyhole limpet haemocyanin; OVA: ovalbumin.

Table 1. Characterization of mouse antiserum against AA-I ($n = 3$).

Mouse	Immunogen	Coating Antigen [1]	Titer	IC_{50} (ng/mL)
1	AA-I-BSA	AA-I-OVA	16K	120.3 ± 13.0
2	AA-I-BSA	AA-I-OVA	32K	148.8 ± 11.8
3	AA-I-BSA	AA-I-OVA	16K	156.1 ± 16.2
4	AA-I-KLH	AA-I-OVA	64K	27.3 ± 4.3
5	AA-I-KLH	AA-I-OVA	64K	56.6 ± 7.7
6	AA-I-KLH	AA-I-OVA	32K	35.8 ± 3.6

[1] The concentration of coating antigen was 1 μg/mL.

3.2. Development of CLEIA

After condition optimization, the CLEIA was further developed. The highest sensitivity was achieved with the coating concentration of 2 μg/mL and antibody concentration of 250 ng/mL (Table S3). For the working buffer, the optimized solution was pH 5.4 PBS with a concentration of 0.01 M (Figure S3). Finally, calibration curves against AA-I were developed and showed the IC_{50} of 2.4 ng/mL with a linear range (IC_{20}–IC_{80}) from 0.2 to 3.1 ng/mL (Figure 2). The IC_{10} (10% decrease in the RLU_{max}) was defined as the limit of detection (LOD), 0.4 ng/mL.

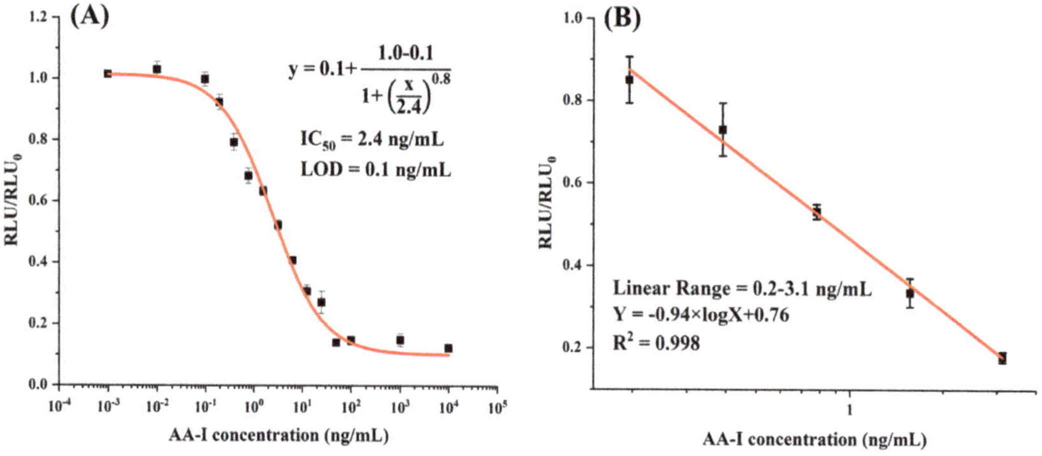

Figure 2. (**A**) The calibration curve; (**B**) the linear range of CLEIA.

Based on the developed CLEIA, the specific test for mAb was performed, and the results are summarized in Table 2. To be noticed, the AA-II showed the highest cross-reactivity (CR) amount of these analogs while only slight CR was observed for AA-III and AA-IV, suggesting the hydroxyl was the key site for recognition. For the other analogs, no obvious CR was observed, indicating the good specificity of mAb to AAs. Since both AA-I and AA-II were the main components in the samples, the obtained mAb can be used for screening these two AAs.

Table 2. The specific test of anti-AA-I mAb.

Compounds	Structure	IC_{50} (ng/mL)	CR [1] (%)
AA-I	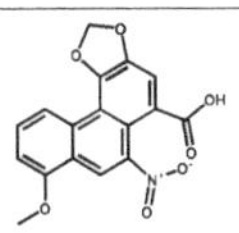	1.8	100
AA-II	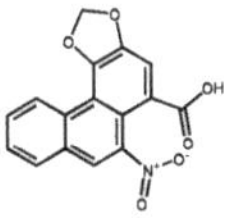	2.1	86
AA-III	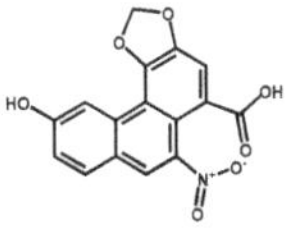	120.0	1.5
AA-IV	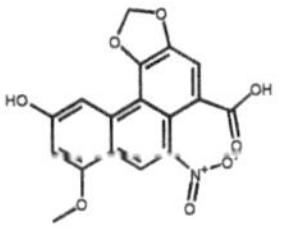	450.0	0.4
Abietic acid	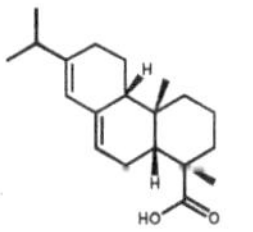	>1000	<0.1
Asarinin	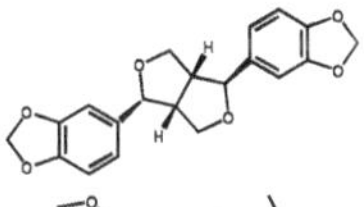	>1000	<0.1
Colchicine	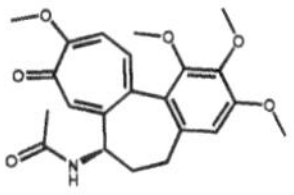	>1000	<0.1
Methyleugenol	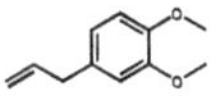	>1000	<0.1
Ephedrine hydrochloride	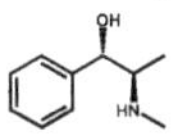	>1000	<0.1

[1] CR(%) = [IC_{50} (AA-I)/IC_{50} (analogues)] × 100%.

3.3. Characterization of rCDs

The rCDs were synthesized and characterized for the development of FIA. As shown in Figure 3A, rCDs exhibited a particle size of approximately 3 nm. Moreover, the high-resolution TEM image clearly showed the lattice fringe of rCDs with an interlayer spacing of 0.21 nm (Figure 3A). The dynamic light scattering (Figure 3B) showed the hydrodynamic size of ~2.7 nm for rCDs, which agreed with TEM.

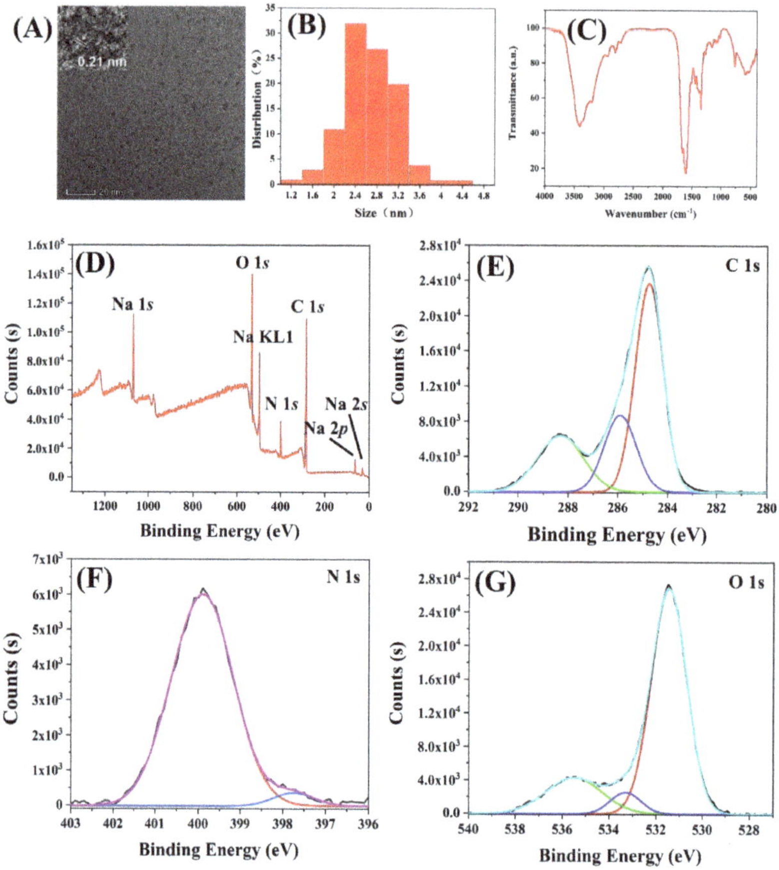

Figure 3. Characterization of rCDs; (**A**) the transmission electron microscope image; (**B**) the hydrodynamic size; (**C**) FTIR image; (**D**) survey XPS spectrum and high resolution of (**E**) C 1s, (**F**) N 1s, and (**G**) O 1s.

The Fourier transform infrared (FTIR) was employed to characterize the functional group of rCDs (Figure 3C). A wide peak from 3100 to 3500 cm^{-1} was found from the FTIR image, which was assigned to the stretching vibrations of the O-H or N-H group, while those from 2800 to 3000 cm^{-1} could be attributed to the stretching vibrations of C-H. The peaks at 1610, 1510, and 1450 cm^{-1} indicated the generation of the aromatic group. The stretching vibrations of C=O or C=N can be verified from the peak of 1680 cm^{-1}, and the peak at 1350 cm^{-1} was caused by the stretching vibrations of C-N, which demonstrated the generation of amide or carboxyl. The peak at 1000–1270 cm^{-1} was originated from stretching vibrations of C-O, implying the existence of ether group or hydroxyl. Furthermore, the peak at 764 cm^{-1} was ascribed to the out-of-plane bending vibration of O-H or C-H demonstrated the ring-shaped conjugated structures in rCDs.

To verify the inference described above, the survey X-ray photoelectron spectroscopy (XPS) was employed. Figure 3D clearly shows the binding energy peaks of C 1 s, N 1s, and O 1s, while the peaks of sodium of Na 1s, Na 2s, Na 2p, and Na KL1 were ascribed to the addition of NaOH in the synthesis of rCDs. The high-resolution XPS showed the three peaks of C 1s (Figure 3E), corresponding to 284.8 eV (C=C/C-C), 285.9 eV (C-N/C-O/C=N), and 288.3 eV(O=C-O). Two fitting peaks of N 1s at 399.9 eV and 397.7 eV were attributed to the pyridinic N and pyrrolic N, respectively (Figure 3F). Figure 3G confirms the presence of C=O (531.5 eV) bonds, O=C-O group (533.5 eV), and the C-O of the aromatic nucleus (535.5 eV). In general, the results of XPS showed good agreement with the generation of the aromatic group from FTIR analysis.

3.4. Development of FIA

The spectral characteristic of rCDs, TMB, and oxTMB was investigated to assess the feasibility of FIA development. Figure 4A shows the Em wavelength of rCDs at 620 nm with the Ex wavelength at 540 nm. Compared with TMB, the oxTMB exhibited an obvious absorbance peak at 650 nm, which overlaps the Em of rCDs, thereby quenching the fluorescent signal of rCDs. Therefore, the presence of HRP can catalyze TMB into oxTMB to quench rCDs; otherwise, the fluorescent signal was a turn-on. The fluorescence lifetime of rCDs was investigated. In the presence of oxTMB, the fluorescence lifetime of rCDs (5.57 ns) showed no obvious difference to rCDs without oxTMB (5.59 ns), indicating the inner-filter effect caused the quenching (Figure 4B). The fluorescent intensity of rCDs with various pH values was also studied, and the highest intensity was observed for the pH value of 11 (Figure S4). Therefore, rCDs were diluted by NaOH (pH 11) to adjust the pH value before adding to microplates.

Based on the quenching mechanism, an rCDs-based FIA was developed, strategy diagram is shown in Figure 4C. In the absence of AA-I, the mAb was bound to coating antigen, resulting in the generation of oxTMB, leading to the quenching of rCDs. In contrast, the presence of AA-I inhibited the binding of mAb to coating antigen, recovering the fluorescent signal of rCDs. Hence AA-I concentration regulated the fluorescence response. Based on the above optimized conditions, a fluorescent calibration curve against AA-I showed the IC_{50} of 0.41 ng/mL (Figure 4D), the LOD of 0.06 ng/mL, and a linear range from 0.08 to 2.5 ng/mL. Compared with other publications for AA-I analysis, the developed FIA showed higher sensitivity (LOD) (Table 3), simplicity, and high efficiency without a complicated procedure for sensitivity improvement, which makes the FIA suitable for sample screening. The downsides are that the FIA still requires at least approximately 90 min to detect AA-I, and the microplate reader requirement limits the on-site detection of FIA. To overcome these shortcomings, handheld reader-based lateral flow immunoassays will be developed in future work for more rapid and on-site detection of AA-I.

Table 3. The comparison of immunoassay for AAs.

Method	IC_{50} (ng/mL)	Linear Range (ng/mL)	LOD (ng/mL)	Reference
icELISA	1.2	ND [1]	0.1	1
Injection analysis chemiluminescence	ND [1]	10–20000	3	4
CLEIA	2.4	0.2–3.1	0.1	This work
FIA	0.41	0.08–2.50	0.06	This work

[1] ND, no data.

3.5. Recovery Test

Three levels of AA-I were spiked to drug homologous and foods then analyzed by CLEIA and FIA. The results were verified by UPLC-QQQ-MS/MS, which is shown in Table 4. The recoveries of CLEIA and FIA were ranged between 83–119% and 86–118.4% with a coefficient of variance (CV) ranging from 3.8% to 13.3% and 5% to 14.4%, respectively. The UPLC-QQQ-MS/MS showed the average recoveries from 85.1% to 108.1%, with CVs from 1.2% to 7.5%. The developed CLEIA and FIA showed good agreement to standard instrument methods, demonstrating good accuracy and practicability for AA-I detection.

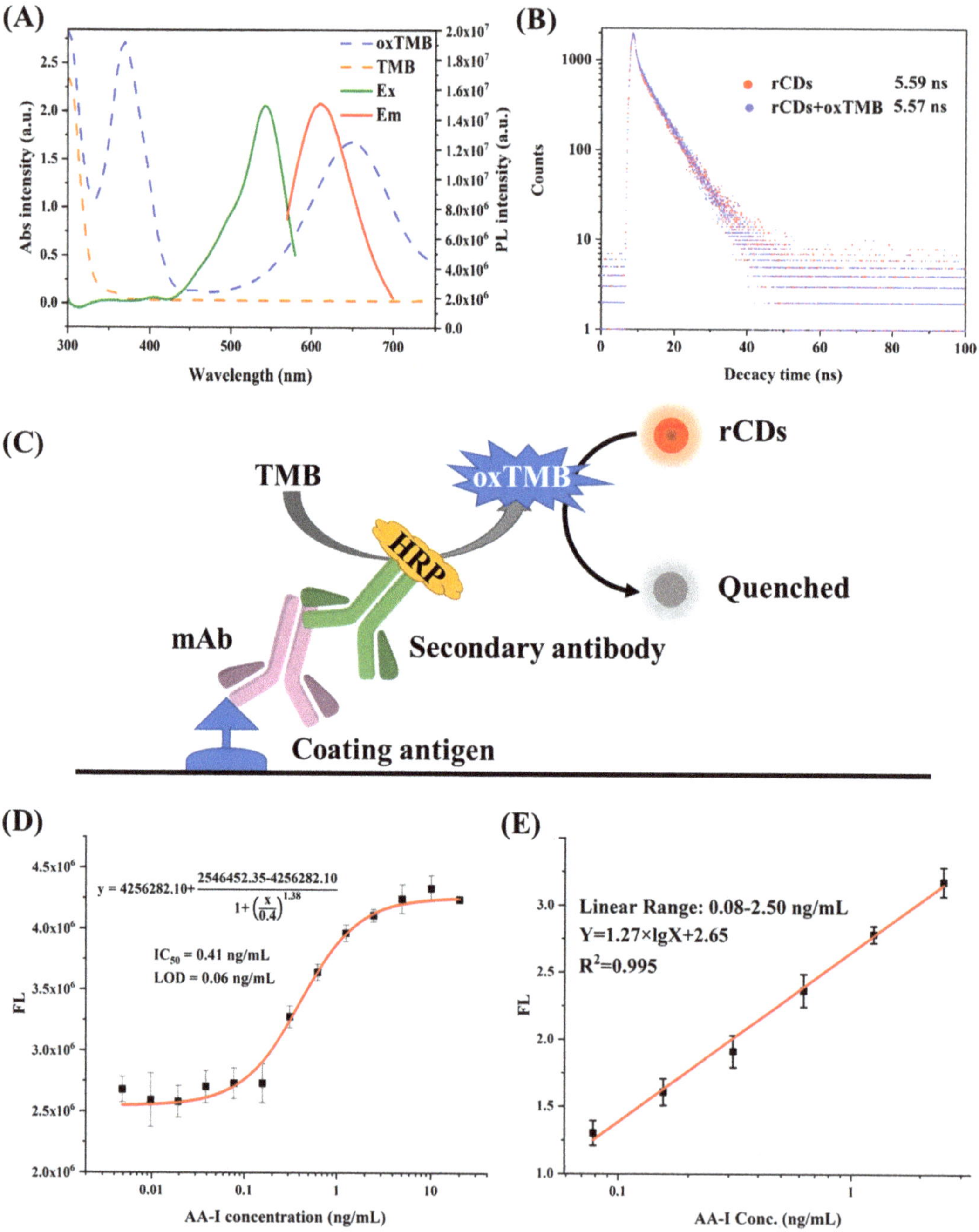

Figure 4. (**A**) The spectrum characterization of rCDs, TMB, and oxTMB. (**B**) The lifetime analysis of rCDs with and without oxTMB. (**C**) Strategic schema of development of FIA. (**D**) Calibration curve and (**E**) linear range of FIA. Acronym: TMB: 3,3′′,5,5′′-tetramethylbenzidine dihydrochloride; oxTMB: oxidized TMB; Em: emission; Ex: excitation.

Table 4. Recovery test for CLEIA, FIA, and UPLC-QQQ-MS/MS (n = 3).

Sample No.	Spiked (ng/g)	CLEIA			FIA			UPLC-QQQ-MS/MS		
		Measured (ng/mL) (Mean ± SD [1])	Recovery (%)	CV [2] (%)	Measured (ng/mL) (Mean ± SD)	Recovery (%)	CV (%)	Measured (ng/mL) (Mean ± SD)	Recovery (%)	CV (%)
Pleurotus ostreatus	600	511.4 ± 41.4	86.7	7.7	655.2 ± 92.4	109.2	14.1	648.3 ± 11.9	108.1	1.8
	300	165.3 ± 21.9	83.0	13.3	355.2 ± 55.2	118.4	15.5	303.7 ± 12.9	101.2	4.2
	150	135.1 ± 14	90.0	12.8	129 ± 18	86	14.0	127.7 ± 9.6	85.1	7.5
Slimming capsule	600	520.1 ± 11.3	86.7	10.4	642 ± 92.4	107	14.4	571.3 ± 12.9	95.2	2.3
	300	237.9 ± 30.4	119.0	12.6	306 ± 28.8	102	9.4	313.7 ± 16.5	104.6	5.3
	150	116 ± 41.1	116.0	36.2	147 ± 21	98	14.3	136 ± 5.6	90.7	4.1
Slimming tea	600	517.4 ± 11.8	86.7	3.8	678 ± 62.4	113	9.2	644.3 ± 7.5	107.4	1.2
	300	204.9 ± 45.7	102.0	22.5	312 ± 15.6	104	5.0	305.3 ± 11.9	101.8	3.9
	150	114.3 ± 13.3	114.0	12.3	138 ± 9	92	6.5	155.3 ± 6.5	103.5	4.2

[1] SD, standard deviation; [2] CV, coefficient of variance.

4. Conclusions

In this study, the AA-I was conjugated to a carrier protein to prepare artificial antigen, and a sensitive anti-AA-I mAb was generated after animal immunization and hybridoma screening. The obtained mAb was further used to develop CLEIA for the detection of AA-I. Since products containing AAs are prohibited in most countries, high sensitivity methods are needed to screen out positive samples, but the sensitivity of CLEIA was still not satisfying. Therefore, CDs with red emission were synthesized and employed to develop FIA, which exhibited a five-fold improvement in sensitivity than CLEIA. The accuracy and practicability of CLEIA and FIA were verified by the standard instrument method; they were sensitive, rapid, and easy to use, making them effective tools for screening AA-I in related products.

Supplementary Materials: The following are available online at https://www.mdpi.com/article/10.3390/foods10112647/s1. Figure S1: Ethical review of animal experiments. Figure S2: Ultraviolet scanning of synthesized antigens of AA-I (A) Ultraviolet spectrum of AA-I, KLH and AA-I-KLH; (B) AA-I, BSA and AA-I-BSA; (C) AA-I, OVA and AA-I-OVA. Figure S3: The optimizing of (A,B) ionic strength; (C,D) pH value; (E,F) Tween-20 concentration; (G, H) methanol concentration. Figure S4: The optimizing of pH for rCDs. Table S1: Parameters of UPLC-QQQ-MS/MS. Table S2: Cell lines evaluation (n = 3). Table S3: Optimization of coating antigen concentration (n = 3).

Author Contributions: Conceptualization, methodology, software, and writing—original draft preparation, A.-F.O.; methodology and software, Z.-J.C.; methodology, Y.-F.Z.; methodology, Q.-Y.H.; conceptualization, Z.-L.X.; conceptualization, writing—reviewing and editing, S.-Q.Z. All authors have read and agreed to the published version of the manuscript.

Funding: Please add: This research was funded by Guangzhou Science and Technology Foundation, grant number 201903010034; Natural Resources Science Foundation of Guangdong Province, grant number 2018A030313926; Science and Technology Foundation Key R&D Program of Guangdong Province, grant number 2019B020209009 & 2019B020218009; R&D Program of Guangdong Province Drug Administration grant number 2021TDZ09 & 2021YDZ06.

Institutional Review Board Statement: The animal experiment was carried out in a laboratory with a license for experiment animal, which was conformed to the welfare principle (ethical approval number: 2019054, Figure S1).

Informed Consent Statement: Not applicable.

Data Availability Statement: The datasets used and analyzed during the current study are available from the corresponding author on request.

Conflicts of Interest: The authors declare no conflict of interest.

References

1. Yu, F.Y.; Lin, Y.H.; Su, C.C. A sensitive enzyme-linked immunosorbent assay for detecting carcinogenic aristolochic acid in herbal remedies. *J. Agric. Food Chem.* **2006**, *54*, 2496–2501. [CrossRef] [PubMed]
2. Yeh, Y.H.; Lee, Y.T.; Hsieh, H.S.; Hwang, D.F. Short-term toxicity of aristolochic acid, aristolochic acid-I and aristolochic acid-II in rats. *Food Chem. Toxicol.* **2008**, *46*, 1157–1163. [CrossRef] [PubMed]
3. Lee, T.Y.; Wu, M.L.; Deng, J.F.; Hwang, D.F. High-performance liquid chromatographic determination for aristolochic acid in medicinal plants and slimming products. *J. Chromatogr. B* **2002**, *766*, 169–174. [CrossRef]
4. Oraby, H.F.; Alarfaj, N.A.; El-Tohamy, M.F. Gold nanoparticle-enhanced luminol/ferricyanide chemiluminescence system for aristolochic acid-I detection in medicinal plants and slimming products. *Green Chem. Lett. Rev.* **2017**, *10*, 138–147. [CrossRef]
5. Ioset, J.R.; Raoelison, G.E.; Hostettmann, K. Detection of aristolochic acid in Chinese phytomedicines and dietary supplements used as slimming regimens. *Food Chem. Toxicol.* **2003**, *41*, 29–36. [CrossRef]
6. Nault, J.; Letouzé, E. Mutational processes in hepatocellular carcinoma: The story of aristolochic acid. *Semin. Liver Dis.* **2019**, *39*, 334–340. [CrossRef]
7. Nortier, J.L.; Martinez, M.M.; Schmeiser, H.H.; Arlt, V.M.; Bieler, C.A.; Petein, M.; Depierreux, M.F.; De Pauw, L.; Abramowicz, D.; Vereerstraeten, P.; et al. Urothelial carcinoma associated with the use of a Chinese herb (*Aristolochia fangchi*). *N. Engl. J. Med.* **2000**, *342*, 1686–1692. [CrossRef]
8. Cosyns, J.; Jadoul, M.; Squifflet, J.; van Cangh, P.; van Ypersele de Strihou, C. Urothelial malignancy in nephropathy due to Chinese herbs. *Lancet* **1994**, *344*, 188. [CrossRef]
9. Zhang, H.M.; Zhao, X.H.; Sun, Z.H.; Li, G.C.; Liu, G.C.; Sun, L.R.; Hou, J.Q.; Zhou, W. Recognition of the toxicity of aristolochic acid. *J. Clin. Pharm. Ther.* **2019**, *44*, 157–162. [CrossRef]
10. Jadot, I.; Declèves, A.; Nortier, J.; Caron, N. An integrated view of aristolochic acid nephropathy: Update of the literature. *Int. J. Mol. Sci.* **2017**, *18*, 297. [CrossRef]
11. Kocic, G.; Gajic, M.; Tomovic, K.; Hadzi Djokic, J.; Anderluh, M.; Smelcerovic, A. Purine adducts as a presumable missing link for aristolochic acid nephropathy-related cellular energy crisis, potential anti-fibrotic prevention and treatment. *Brit. J. Pharmacol.* **2021**, *178*, 4411–4427.
12. Koyama, N.; Yonezawa, Y.; Nakamura, M.; Sanada, H. Evaluation for a mutagenicity of aristolochic acid by Pig-a and PIGRET assays in rats. *Mutat. Res. Genet. Toxicol. Environ. Mutagen. Environ. Mutagenesis* **2016**, *811*, 80–85. [CrossRef]
13. Vanherweghem, J.; Tielemans, C.; Abramowicz, D.; Depierreux, M.; Vanhaelen-Fastre, R.; Vanhaelen, M.; Dratwa, M.; Richard, C.; Vandervelde, D.; Verbeelen, D.; et al. Rapidly progressive interstitial renal fibrosis in young women: Association with slimming regimen including Chinese herbs. *Lancet* **1993**, *341*, 387–391. [CrossRef]
14. Depierreux, M.; Van Damme, B.; Vanden Houte, K.; Vanherweghem, J.L. Pathologic aspects of a newly described nephropathy related to the prolonged use of Chinese herbs. *Am. J. Kidney Dis.* **1994**, *24*, 172–180. [CrossRef]
15. Koh, H.L.; Wang, H.; Zhou, S.; Chan, E.; Woo, S.O. Detection of aristolochic acid I, tetrandrine and fangchinoline in medicinal plants by high performance liquid chromatography and liquid chromatography/mass spectrometry. *J. Pharmaceut. Biomed.* **2006**, *40*, 653–661. [CrossRef]
16. Guo, L.; Yue, H.; Cai, Z.W. A novel pre-column fluorescent derivatization method for the sensitive determination of aristolochic acids in medicinal herbs by high-performance liquid chromatography with fluorescence detection. *J. Pharmaceut. Biomed.* **2010**, *53*, 37–42. [CrossRef]
17. Wang, Y.A.; Chan, W. Determination of aristolochic acids by high-performance liquid chromatography with fluorescence detection. *J. Agric. Food Chem.* **2014**, *62*, 5859–5864. [CrossRef]
18. Wang, X.R.; Wang, Y.Y.; Wang, Y.D.; Chen, Q.; Liu, X. Nanobody-alkaline phosphatase fusion-mediated phosphate-triggered fluorescence immunoassay for ochratoxin a detection. *Spectrochim. Acta A* **2020**, *226*, 117617. [CrossRef]
19. Inui, H.; Takeuchi, T.; Uesugi, A.; Doi, F.; Takai, M.; Nishi, K.; Miyake, S.; Ohkawa, H. Enzyme-linked immunosorbent assay with monoclonal and single-chain variable fragment antibodies Selective to Coplanar Polychlorinated Biphenyls. *J. Agric. Food Chem.* **2012**, *60*, 1605–1612. [CrossRef]
20. Zhou, J.J.; Ren, M.S.; Wang, W.J.; Huang, L.; Lu, Z.C.; Song, Z.Y.; Foda, M.F.; Zhao, L.; Han, H.Y. Pomegranate-inspired silica nanotags enable sensitive dual-modal detection of rabies virus nucleoprotein. *Anal. Chem.* **2020**, *92*, 8802–8809. [CrossRef]
21. Liu, Y.Z.; Zhao, G.X.; Wang, P.; Liu, J.; Zhang, H.C.; Wang, J.P. Production of the broad specific monoclonal antibody against sarafloxacin for rapid immunoscreening of 12 fluoroquinolones in meat. *J. Environ. Sci. Heal. B* **2013**, *48*, 139–146. [CrossRef]
22. Li, Y.; Liu, L.Q.; Kuang, H.; Xu, C.L. Preparing monoclonal antibodies and developing immunochromatographic strips for paraquat determination in water. *Food Chem.* **2020**, *311*, 125897.1–125897.9. [CrossRef]
23. Wu, Y.P.; Wang, J.; Zhou, Y.; Qi, Y.; Ma, L.C.; Wang, X.N.; Tao, X.Q. Quantitative determination of nitrofurazone metabolites in animal-derived foods based on a background fluorescence quenching immunochromatographic Assay. *Foods* **2021**, *10*, 1668. [CrossRef]
24. Chen, X.R.; Miao, X.T.; Ma, T.T.; Leng, Y.K.; Hao, L.W.; Duan, H.; Yuan, J.; Li, Y.; Huang, X.L.; Xiong, Y.H. Gold nanobeads with enhanced absorbance for improved sensitivity in competitive lateral flow immunoassays. *Foods* **2021**, *10*, 1488. [CrossRef]
25. Tao, X.Q.; Zhou, S.; Yuan, X.M.; Li, H.J. Determination of chloramphenicol in milk by ten chemiluminescent immunoassays: Influence of assay format applied. *Anal. Methods* **2016**, *8*, 4445–4451. [CrossRef]

26. Xu, L.; Suo, X.Y.; Zhang, Q.; Li, X.P.; Chen, C.; Zhang, X.Y. ELISA and chemiluminescent enzyme immunoassay for sensitive and specific determination of lead (II) in Water, Food and Feed Samples. *Foods* **2020**, *9*, 305. [CrossRef]
27. Yi, K.Y.; Zhang, X.T.; Zhang, L. Eu^{3+}@metal–organic frameworks encapsulating carbon dots as ratiometric fluorescent probes for rapid recognition of anthrax spore biomarker. *Sci. Total Environ.* **2020**, *743*, 140692. [CrossRef]
28. Li, T.X.; Li, Z.; Huang, T.Z.; Tian, L. Carbon quantum dot-based sensors for food safety. *Sens. Actuators A Phys.* **2021**, *331*, 113003. [CrossRef]
29. Song, P.; Liu, Q.; Zhang, Y.; Liu, W.; Meng, M.; Yin, Y.M.; Xi, R.M. The chemical redox modulated switch-on fluorescence of carbon dots for probing alkaline phosphatase and its application in an immunoassay. *RSC Adv.* **2018**, *8*, 162–169. [CrossRef]
30. Ni, P.J.; Xie, J.F.; Chen, C.X.; Jiang, Y.Y.; Lu, Y.Z.; Hu, X. Fluorometric determination of the activity of alkaline phosphatase and its inhibitors based on ascorbic acid-induced aggregation of carbon dots. *Microchim. Acta* **2019**, *202*, 186. [CrossRef]
31. Li, G.L.; Fu, H.L.; Chen, X.J.; Gong, P.W.; Chen, G.; Xia, L.; Wang, H.; You, J.M.; Wu, Y.N. Facile and sensitive fluorescence sensing of alkaline phosphatase activity with photoluminescent carbon dots based on inner filter effect. *Anal. Chem.* **2016**, *88*, 2720–2726. [CrossRef] [PubMed]
32. Chen, Z.J.; Liu, X.X.; Xiao, Z.L.; Fu, H.J.; Huang, Y.P.; Huang, S.Y.; Shen, Y.D.; He, F.; Yang, X.X.; Hammock, B.; et al. Production of a specific monoclonal antibody for 1-naphthol based on novel hapten strategy and development of an easy-to-use ELISA in urine samples. *Ecotox. Environ. Saf.* **2020**, *196*, 110533. [CrossRef] [PubMed]
33. Chen, Z.J.; Zhou, K.; Ha, W.Z.; Chen, P.H.; Fu, H.J.; Shen, Y.D.; Sun, Y.M.; Xu, Z.L. Development of a low-cost, simple, fast and quantitative lateral-flow immunochromatographic assay (ICA) strip for melatonin in health foods. *Food Agric. Immunol.* **2019**, *30*, 497–509. [CrossRef]

foods

Article

Influence of Endogenous Factors of Food Matrices on Avidin—Biotin Immunoassays for the Detection of Bacitracin and Colistin in Food

Maksim A. Burkin [1,*], Inna A. Galvidis [1] and Sergei A. Eremin [2,*]

[1] Immunology Department, I. Mechnikov Research Institute for Vaccines and Sera, 105064 Moscow, Russia; galvidis@yandex.ru
[2] Faculty of Chemistry, M. V. Lomonosov Moscow State University, Leninsky Gory, 1, 119991 Moscow, Russia
* Correspondence: burma68@yandex.ru (M.A.B.); eremin_sergei@hotmail.com (S.A.E.); Tel.: +7-495-9172753 (M.A.B.)

Abstract: (Strept)avidin–biotin technology is frequently used in immunoassay systems to improve their analytical properties. It is known from clinical practice that many (strept)avidin–biotin-based tests provide false results when analyzing patient samples with a high content of endogenous biotin. No specific investigation has been carried out regarding possible interferences from avidin (AVI) and biotin (B_7) contained in food matrices in (strept)avidin–biotin-based immunoanalytical systems for food safety. Two kinds of competitive ELISAs for bacitracin (BT) and colistin (COL) determination in food matrices were developed based on conventional hapten–protein coating conjugates and biotinylated BT and COL bound to immobilized streptavidin (SAV). Coating SAV–B_7–BT and SAV–B_7–COL complexes-based ELISAs provided 2- and 15-times better sensitivity in BT and COL determination, corresponding to 0.6 and 0.3 ng/mL, respectively. Simultaneously with the determination of the main analytes, these kinds of tests were used as competitive assays for the assessment of AVI or B_7 content up to 10 and 1 ng/mL, respectively, in food matrices (egg, infant milk formulas enriched with B_7, chicken and beef liver). Matrix-free experiments with AVI/B_7-enriched solutions showed distortion of the standard curves, indicating that these ingredients interfere with the adequate quantification of analytes. Summarizing the experience of the present study, it is recommended to avoid immunoassays based on avidin–biotin interactions when analyzing biosamples containing these endogenous factors or enriched with B_7.

Keywords: bacitracin; colistin; biotinylated hapten coating; immunoassay; matrix biotin and avidin interference; food contaminants

Citation: Burkin, M.A.; Galvidis, I.A.; Eremin, S.A. Influence of Endogenous Factors of Food Matrices on Avidin—Biotin Immunoassays for the Detection of Bacitracin and Colistin in Food. *Foods* **2022**, *11*, 219. https://doi.org/10.3390/foods11020219

Academic Editor: Simon Haughey

Received: 18 December 2021
Accepted: 12 January 2022
Published: 13 January 2022

Publisher's Note: MDPI stays neutral with regard to jurisdictional claims in published maps and institutional affiliations.

1. Introduction

The interaction of (strept)avidin protein and biotin (vitamin B_7) is widely used in various bioanalytical systems, in particular, in immunoassays [1]. Due to the high affinity ($K_D \approx 10^{-15}$ M) and binding valence between this protein (four binding sites) and the vitamin, they are successfully applied as immunoreagent labels to provide an additional functionality [2], for oriented immobilization/presentation [3], as well as to accelerate the interaction [4] or enhance the output signal in immunoassays of various designs [5,6]. In this regard, the involvement of the (strept)avidin–biotin system is one of the main strategies for increasing the sensitivity of immunoassay [7].

In a wide panel of commercial (strept)avidin–biotin-based immunoassay systems used in clinical practice for diagnostic purposes, it has been found that the presence of endogenous biotin in patients' biofluids can interfere with the analysis, leading to false results and, consequently, to misdiagnosis and patient mismanagement [8,9]. However, with a dietary intake of about 35–70 µg B_7/day, the blood B_7 level in healthy subjects (0.12–0.36 nM) [10] has a negligible interfering effect on (strept)avidin–biotin assays. At the same time, excessive consumption as a result of therapy for a number of disorders (multiple sclerosis, phenylketonuria, biotinidase deficiency) or incorporation in "Hair, Skin, and Nails" cosmetic formulas may lead to μM B_7 blood levels, which can significantly distort the results of (strept)avidin–biotin-based tests [9].

The (strept)avidin–biotin technology is equally popular in immunoassays for food safety control [11–13]. However, researchers have not yet been concerned about the possibility that avidin and biotin in the matrix could interfere with the detection of a wide range of analytes in food. Avidin (AVI) is a known protein component in egg white, whereas biotin is widely distributed in different natural foodstuffs. Foods relatively rich in biotin include egg yolk, liver, some vegetables [14], and the majority of biotin in meats and cereals appears to be protein-bound [10]. Thus, these endogenous components of the matrix, the content of which reaches ppb–ppm levels in many analyzed natural foods [15], can represent a serious obstacle in immunoassays based on (strept)avidin–biotin interaction.

In this regard, the objective of the present study was the assessment of the influence of these factors on newly developed ELISA systems based on the coating of biotinylated haptens for the determination of the antibiotics bacitracin and colistin in food matrices rich in AVI and B_7 components.

Bacitracin (BT) [MW = 1422.7] and colistin (COL) [MW = 1155.4] are both cyclic peptide antibiotics (Figure 1) used in veterinary and human medicine. BT is produced by strains of *Bacillus licheniformis* and functions as an anti-Gram-positive agent, inhibiting bacterial cell wall biosynthesis [16]. Colistin is a product of the biosynthesis of *Paenibacillus polymyxa*, has a wide Gram-negative spectrum, and kills bacteria mainly through the disruption of bacterial outer membrane integrity, resulting from the binding with LPS [17].

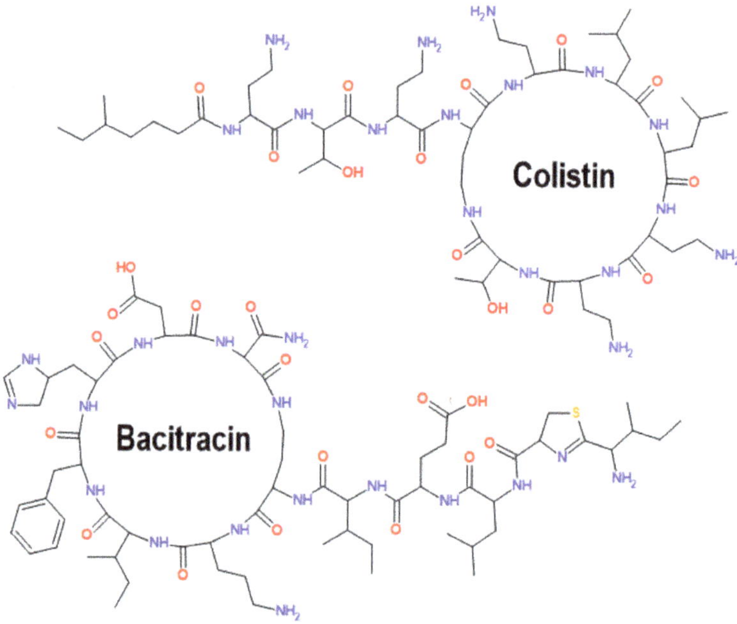

Figure 1. Structural formulas of the peptide antibiotics bacitracin and colistin.

To expand their antibacterial action, these antibiotics can be used together, potentiating each other's activity [18]. Their use in farm animals and a proper withdrawal period should be controlled, so that the residual content of these antibiotics in agricultural products does not exceed the established acceptable threshold (Table 1).

of extraneous AVI on ELISAs in the detection of BT and COL using the coated SAV–B_7–hapten complex was examined. This effect was investigated in different stages of the assay. Figure 4 (dashed lines) demonstrates that extraneous AVI (1–100,000 ng/mL) could inhibit the formation of the complexes SAV–BT–B_7 and SAV–COL–B_7 in a dose-dependent manner (stage 1). Therefore, the developed assay systems could simultaneously serve as competitive assays for AVI detection. No evident effect at stages 2 and 3 was found when SAV–B_7 complexes were already formed. Thus, interference from extraneous AVI may be a drawback for assay formats involving "coated antibody–hapten–B_7" or "coated hapten–antibody–B_7", where AVI can interact with unblocked biotin.

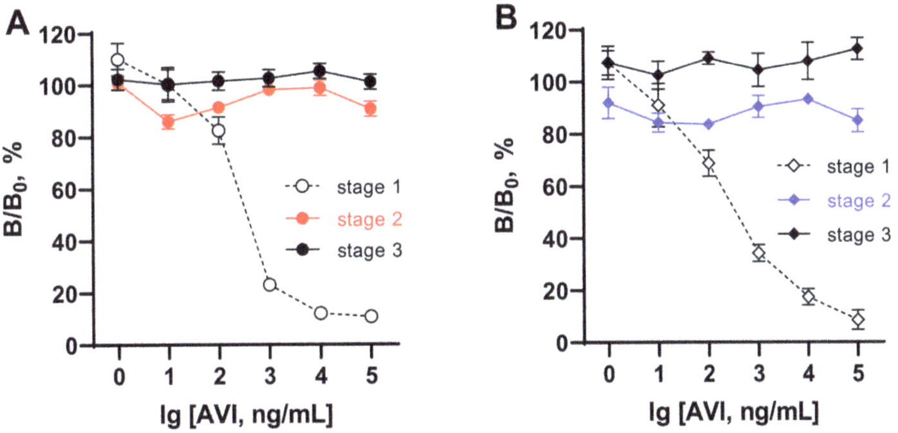

Figure 4. Influence of extraneous AVI on different stages of ELISAs for bacitracin (**A**) and colistin (**B**) based on coated complexes formed between SAV and biotinylated hapten. Complex formation between coated SAV and BT/COL–B_7 (stage 1), antibody–analyte binding (stage 2), and bound antibody–GAR–HRP interaction (stage 3).

3.4. *Assessment of Extraneous Biotin Influence on the Assay*

Mammals cannot synthesize biotin but depend on dietary intake from microbial and plant sources [10]. Nevertheless, B_7 uptake, accumulation in tissues, especially in the liver, and renal reabsorption of B_7 determine its presence in animal-derived food. The tissues richest in biotin are chicken and beef liver (0.4–1.9 μg/g), egg yolk (0.3 μg/g), fish (0.05–0.1 μg/g), while the majority of B_7 in meat is protein-bound [15]. The mentioned B_7 content is sufficient to cause an undesirable interference in (strept)avidin–biotin-based immunoassay.

The possible interference from extraneous B_7 was examined in different stages of the developed assays as reported above for the assessment of AVI influence. As seen in Figure 5, the presence of B_7 in the test samples up to 10,000 ng/mL (50 μM) could disrupt the formation of the complex between SAV and B_7 at concentrations >1 ng/mL (Figure 5, stage 1). When the complexes between SAV and B_7–hapten were already formed in stage 1, the influence of extraneous B_7 could not strongly affect the binding in the assay ranging between 80 and 110% (Figure 5, stage 2,3).

Thus, these experiments showed that samples with moderate to high AVI or B_7 content cannot be analyzed using tests in which the analyte is detected during SAV–B_7 complexation due to the strong influence of endogenous matrix factors. In the assay design considered in this work, the recognition of the analyte by antibodies (stage 2) was separated from the SAV–B_7 interaction (stage 1) to minimize the possible influence of endogenous AVI/B_7 from the matrix.

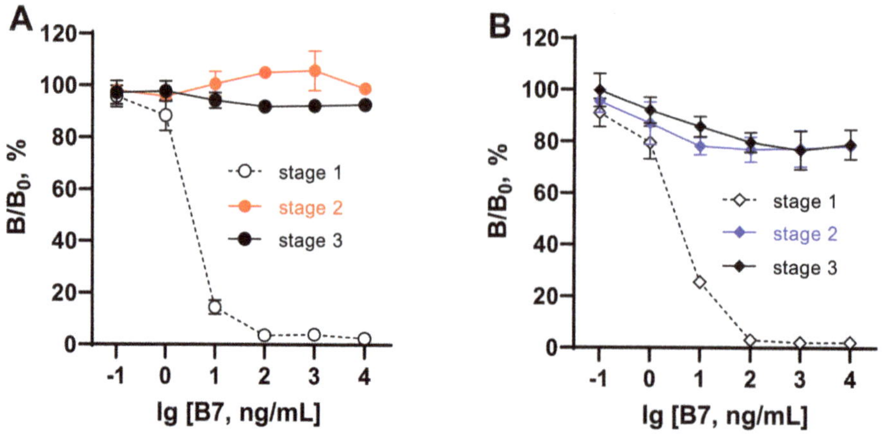

Figure 5. Influence of extraneous biotin (B_7) in different stages of ELISAs for bacitracin (**A**) and colistin (**B**) based on coated complexes formed between SAV and biotinylated hapten. Complex formation between coated SAV and BT/COL–B_7 (stage 1), antibody–analyte binding (stage 2), and bound antibody–GAR–HRP interaction (stage 3).

3.5. Influence Assessment of Avidin and Biotin Components from Different Matrices

Egg. Poultry eggs are real animal-derived samples among possible biotin-binding food matrices, which can be analyzed for antibiotic contaminants such as BT or COL. The maximum concentration of AVI in chicken egg is about 0.05% of the total egg protein (approximately 1800 µg per egg) [33]. B_7 is also found in eggs, about 7 and 50 µg per 100 g of egg white and yolk, respectively (20.7 µg in a whole egg). Thus, both components of the egg are capable of interfering with the analysis. On the other hand, being components of one sample, they can partly quench each other's activity. However, their residual inhibitory activity on SAV–B_7 binding remains unknown. To estimate the inhibitory effect of extraneous AVI/B_7 from eggs on ELISAs for the detection of BT and COL, homogenates were prepared from eggs produced in poultry farms from seven different country regions (Leningrad Oblast, Nizhny Novgorod Oblast, Yaroslavl Oblast, Tula Oblast, Ryazan Oblast, Udmurtia, and Bashkortostan). Egg homogenates, 100-fold diluted with PBST, could inhibit the binding of the B_7-labeled haptens to the SAV-coated wells when they were added together with the latter. Using standard curves (dashed lines from Figures 4 and 5), their resulting activity was measured and corresponded to an AVI average content of 90.6 ± 47.5 µg/mL (20.6–145.5) and to a B_7 average content of 1.98 ± 0.92 µg/mL (0.56–3.0).

Milk formula. Milk itself contains negligible amounts of B_7, about 1 ng/g [15], but some infant formulas are fortified with vitamins. Two infant milk formula, "Agusha-1" and fermented milk formula "Agusha-2", which were used in recovery experiments of BT/COL, included 2 µg B_7 per 100 g of product (20 ng/mL), as indicated by the manufacturer.

Liver. Animal liver is one of the richest sources of biotin [15]. Samples of chicken and beef liver were examined for their interference in the formation of the SAV–B_7–hapten complex and demonstrated inhibitory activity equivalent to 325 ± 25 ng/g and 164 ± 36 ng/g, respectively.

Thus, the considered food matrices include a sufficient level of endogenous factors, AVI or B_7, which can inhibit the binding of SAV- or B_7-labeled reagents and distort the assay results. This was exemplified by the influence of the matrix at the stage of binding of the biotinylated hapten to the immobilized SAV. However, despite the stage of analyte recognition and the stage of SAV–B7 interaction being separated in the developed assay systems, the influence of endogenous matrix factors on the quantification of analytes was tested using standard BT and COL curves.

3.6. Influence of AVI and B_7 on the Quantification of BT and COL with ELISAs Based on the Coated SAV–B_7–Hapten Complex

To simulate the effect of AVI and B_7 contained in a matrix on the quantitative determination of BT and COL, we examined the calibration curves obtained using media enriched with these factors. Analyte standards were prepared in PBST and in buffer with the maximum expected concentration of the studied factors in food samples (AVI, 100,000 ng/mL, B_7, 10–1000 ng/mL). The comparative study of standard curves generated in buffer and media containing AVI (Figure 6) and B_7 (Figure 7) revealed a discrepancy between the compared curves. These experiments with model AVI/B_7-rich food matrices confirmed the interference of these endogenous factors in analyte quantification in (strept)avidin-based assays. This finding provides an explanation for the failure of the recovery of BT and COL from AVI/B_7-rich foodstuffs using these kinds of assays.

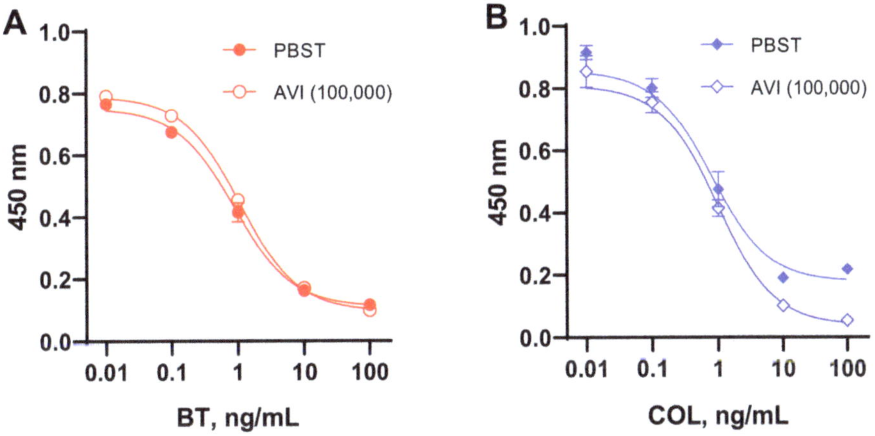

Figure 6. Standard curves for bacitracin (**A**) and colistin (**B**) determination in SAV–B_7–hapten ELISAs generated in PBST and buffer enriched with avidin (AVI) at 100,000 ng/mL.

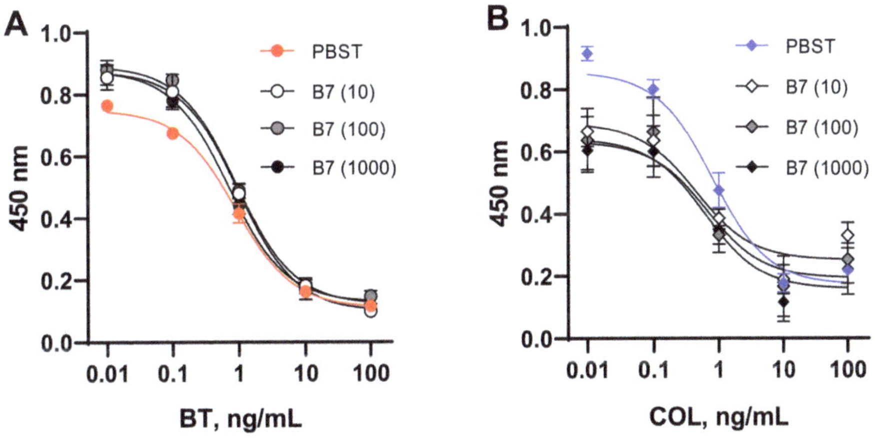

Figure 7. Standard curves for bacitracin (**A**) and colistin (**B**) determination in SAV–B_7–hapten ELISAs generated in PBST and buffer enriched with biotin (B_7) at 10, 100, and 1000 ng/mL.

4. Conclusions

An alternative approach for hapten coating on plates was realized using the oriented binding of biotinylated BT and COL to immobilized SAV. Such formed complexes and BT/COL conjugated to protein carriers were compared as coating antigens in competitive ELISAs for the determination of these cyclic peptide antibiotics in food matrices. The simple titration of B_7–haptens in SAV-coated plates provided a finer optimization of coated hapten load compared with hapten load on the conjugates. Using the coating complexes SAV–B_7–BT and SAV–B_7–COL instead of the coating conjugates GEL–BT(ae) and GEL(pi)–PMB improved the sensitivity of the determination of BT and COL by 2 and 15 times, achieving 0.6 and 0.3 ng/mL, respectively. The applicability of an immunoassay based on (strept)avidin–biotin interactions for the detection of analytes in food matrices rich in AVI or B_7 was also challenged in this work. The possible interference of AVI or B_7 from a matrix in different stages of the immunoassay was investigated. As a result, the developed competitive indirect ELISAs based on coated SAV–B_7–BT and SAV–B–COL, in addition to determining the main analytes, BT and COL, were applied for assessing AVI and B_7 content in food matrices such as eggs, infant milk formula enriched with B_7, and chicken and beef liver. Thus, due to the dose-dependent inhibition of the binding of B_7–hapten to SAV, these ELISAs proved to be suitable for solving additional analytical tasks, detecting AVI and B_7 up to 10 and 1 ng/mL, respectively. The interaction between SAV and B_7–hapten was the one most influenced by extraneous AVI and B_7. The detection of the analyte by an antibody occurs at different stages of the immunoassay, which minimizes the possible interference from AVI and B_7 contained in the matrix. However, an adequate quantitative assessment of BT and COL in food matrices rich in AVI/B_7 was not implemented in the developed (strept)avidin–biotin-based immunoassays. Modification of the standard curves in a matrix-free but AVI/B_7-enriched environment provided evidence of the influence of these factors on the quantification of the analytes. This explains the failure of numerous recovery experiments with fortified nutritional matrices. Thus, summarizing the experience gained in the course of this study, it is recommended to avoid immunoassays based on avidin–biotin interactions when analyzing biosamples containing these factors.

Author Contributions: Conceptualization, Formal Analysis and Investigation, Writing—Original Draft Preparation, Review and Editing, M.A.B.; Formal Analysis and Investigation, Writing—Original Draft Preparation, I.A.G.; Funding Acquisition, Writing—Review and Editing, S.A.E. All authors have read and agreed to the published version of the manuscript.

Funding: This research was supported by the Interdisciplinary Scientific and Educational School of Moscow University «Future Planet and Global Environmental Change».

Institutional Review Board Statement: Not applicable.

Informed Consent Statement: Not applicable.

Data Availability Statement: The datasets used and/or analyzed during the current study are available from the corresponding author on request.

Conflicts of Interest: The authors declare no conflict of interest.

References

1. Wilchek, M.; Bayer, E.A. [54] Avidin-biotin mediated immunoassays: Overview. *Methods Enzymol.* **1990**, *184*, 467–469. [PubMed]
2. El-Maghrabey, M.; Kishikawa, N.; Harada, S.; Ohyama, K.; Kuroda, N. Quinone-based antibody labeling reagent for enzyme-free chemiluminescent immunoassays. Application to avidin and biotinylated anti-rabbit IgG labeling. *Biosens. Bioelectron.* **2020**, *160*, 112215. [CrossRef]
3. Mustafaoglu, N.; Alves, N.J.; Bilgicer, B. Oriented immobilization of Fab fragments by site-specific biotinylation at the conserved nucleotide binding site for enhanced antigen detection. *Langmuir* **2015**, *31*, 9728–9736. [CrossRef] [PubMed]
4. Ohashi, T.; Mawatari, K.; Kitamori, T. On-chip antibody immobilization for on-demand and rapid immunoassay on a microfluidic chip. *Biomicrofluidics* **2010**, *4*, 032207. [CrossRef] [PubMed]
5. Bratthauer, G.L. The avidin–biotin complex (ABC) method and other avidin–biotin binding methods. In *Immunocytochemical Methods and Protocols*; Springer: Berlin/Heidelberg, Germany, 2010; pp. 257–270.

6. Cho, I.-H.; Bhunia, A.; Irudayaraj, J. Rapid pathogen detection by lateral-flow immunochromatographic assay with gold nanoparticle-assisted enzyme signal amplification. *Int. J. Food Microbiol.* **2015**, *206*, 60–66. [CrossRef]
7. Cohen, L.; Walt, D.R. Evaluation of antibody biotinylation approaches for enhanced sensitivity of single molecule array (Simoa) immunoassays. *Bioconjugate Chem.* **2018**, *29*, 3452–3458. [CrossRef] [PubMed]
8. Li, J.; Wagar, E.A.; Meng, Q.H. Comprehensive assessment of biotin interference in immunoassays. *Clin. Chim. Acta* **2018**, *487*, 293–298. [CrossRef] [PubMed]
9. Luong, J.H.; Male, K.B.; Glennon, J.D. Biotin interference in immunoassays based on biotin-strept (avidin) chemistry: An emerging threat. *Biotechnol. Adv.* **2019**, *37*, 634–641. [CrossRef]
10. Zempleni, J.; Wijeratne, S.S.; Hassan, Y.I. Biotin. *Biofactors* **2009**, *35*, 36–46. [CrossRef]
11. Jiang, W.; Beier, R.C.; Luo, P.; Zhai, P.; Wu, N.; Lin, G.; Wang, X.; Xu, G. Analysis of pirlimycin residues in beef muscle, milk, and honey by a biotin–streptavidin-amplified enzyme-linked immunosorbent assay. *J. Agric. Food Chem.* **2016**, *64*, 364–370. [CrossRef]
12. Sun, Z.; Lv, J.; Liu, X.; Tang, Z.; Wang, X.; Xu, Y.; Hammock, B.D. Development of a nanobody-aviTag fusion protein and its application in a streptavidin–biotin-amplified enzyme-linked immunosorbent assay for ochratoxin A in cereal. *Anal. Chem.* **2018**, *90*, 10628–10634. [CrossRef]
13. Lu, M.; Liang, M.; Pan, J.; Zhong, Y.; Zhang, C.; Cui, X.; Wang, T.; Yan, J.; Ding, J.; Zhao, S. Development of a Highly Sensitive Biotin-Streptavidin Amplified Enzyme-Linked Immunosorbent Assay for Determination of Progesterone in Milk Samples. *Food Anal. Methods* 2021. [CrossRef]
14. Zempleni, J.; Mock, D. Biotin biochemistry and human requirements. *J. Nutr. Biochem.* **1999**, *10*, 128–138. [CrossRef]
15. Staggs, C.; Sealey, W.; McCabe, B.; Teague, A.; Mock, D. Determination of the biotin content of select foods using accurate and sensitive HPLC/avidin binding. *J. Food Compos. Anal.* **2004**, *17*, 767–776. [CrossRef] [PubMed]
16. Stone, K.J.; Strominger, J.L. Mechanism of action of bacitracin: Complexation with metal ion and C55-isoprenyl pyrophosphate. *Proc. Natl. Acad. Sci. USA* **1971**, *68*, 3223–3227. [CrossRef] [PubMed]
17. Warren, H.S.; Kania, S.A.; Siber, G. Binding and neutralization of bacterial lipopolysaccharide by colistin nonapeptide. *Antimicrob. Agents Chemother.* **1985**, *28*, 107–112. [CrossRef] [PubMed]
18. Si, W.; Wang, L.; Usongo, V.; Zhao, X. Colistin induces S. aureus susceptibility to bacitracin. *Front. Microbiol.* **2018**, *9*, 2805. [CrossRef] [PubMed]
19. Regulation, E.C. EU Council Regulation (2010) N 37/2010 of 22 December 2009 on pharmacologically active substances and their classification regardingmaximum residue limits in foodstuffs of animal origin. *Off. J. Eur. Union* **2010**, *L15*, 1–72.
20. Hygienic Requirements in Respect of the Safety and Nutritional Value of Foodstuffs, SanPiN 2.3.2.1078-01, as Amended. Chapter 1. Requirements for the Safety and Nutrition of Foods of the Unified Sanitary-Epidemiological and Hygiene Requirements of the Commission of the Customs Union of Russia, Belarus and Kazakhstan. 2019. Available online: http://docs.cntd.ru/document/902249109 (accessed on 10 December 2021).
21. Gaugain, M.; Raynaud, A.; Bourcier, S.; Verdon, E.; Hurtaud-Pessel, D. Development of a liquid chromatography-tandem mass spectrometry method to determine colistin, bacitracin and virginiamycin M1 at cross-contamination levels in animal feed. *Food Addit. Contam. Part A* **2021**, *38*, 1481–1494. [CrossRef]
22. Gaudin, V.; Hédou, C.; Rault, A.; Verdon, E.; Soumet, C. Evaluation of three ELISA kits for the screening of colistin residue in porcine and poultry muscle according to the European guideline for the validation of screening methods. *Food Addit. Contam. Part A* **2020**, *37*, 1651–1666. [CrossRef]
23. Li, Y.; Jin, G.; Liu, L.; Kuang, H.; Xiao, J.; Xu, C. A portable fluorescent microsphere-based lateral flow immunosensor for the simultaneous detection of colistin and bacitracin in milk. *Analyst* **2020**, *145*, 7884–7892. [CrossRef] [PubMed]
24. Byzova, N.A.; Serchenya, T.S.; Vashkevich, I.I.; Zherdev, A.V.; Sviridov, O.V.; Dzantiev, B.B. Lateral flow immunoassay for rapid qualitative and quantitative control of the veterinary drug bacitracin in milk. *Microchem. J.* **2020**, *156*, 104884. [CrossRef]
25. Galvidis, I.A.; Eremin, S.A.; Burkin, M.A. Development of indirect competitive enzyme-linked immunoassay of colistin for milk and egg analysis. *Food Agric. Immunol.* **2020**, *31*, 424–434. [CrossRef]
26. Kononenko, G.; Burkin, A. Methods of sanitary surveillance for livestock production. II. Enzyme immunoassay (EIA) of bacitracin. *Agric. Biol.* 2010; 6, 88–93.
27. Steinbuch, M.; Audran, R. The isolation of IgG from mammalian sera with the aid of caprylic acid. *Arch. Biochem. Biophys.* **1969**, *134*, 279–284. [CrossRef]
28. Galvidis, I.A.; Burkin, K.M.; Eremin, S.A.; Burkin, M.A. Group-specific detection of 2-deoxystreptamine aminoglycosides in honey based on antibodies against ribostamycin. *Anal. Methods* **2019**, *11*, 4620–4628. [CrossRef]
29. Burkin, M.; Galvidis, I. Simultaneous and differential determination of drugs and metabolites using the same antibody: Difloxacin and sarafloxacin case. *Anal. Methods* **2016**, *8*, 5843–5850. [CrossRef]
30. Burkin, M.; Galvidis, I. Immunochemical detection of apramycin as a contaminant in tissues of edible animals. *Food Control* **2013**, *34*, 408–413. [CrossRef]
31. Burkin, M.; Galvidis, I. Development and application of indirect competitive enzyme immunoassay for detection of neomycin in milk. *Appl. Biochem. Microbiol.* **2011**, *47*, 321–326. [CrossRef]
32. Burkin, M.A.; Galvidis, I.A.; Surovoy, Y.A.; Plyushchenko, I.V.; Rodin, I.A.; Tsarenko, S.V. Development of ELISA formats for polymyxin B monitoring in serum of critically ill patients. *J. Pharm. Biomed. Anal.* **2021**, *204*, 114275. [CrossRef]
33. Green, N.M. Avidin. In *Advances in Protein Chemistry*; Elsevier: Amsterdam, The Netherlands, 1975; Volume 29, pp. 85–133.

34. Taskinen, B.; Zmurko, J.; Ojanen, M.; Kukkurainen, S.; Parthiban, M.; Määttä, J.A.; Leppiniemi, J.; Jänis, J.; Parikka, M.; Turpeinen, H. Zebavidin-An avidin-like protein from zebrafish. *PLoS ONE* **2013**, *8*, e77207. [CrossRef]
35. Määttä, J.A.; Helppolainen, S.H.; Hytönen, V.P.; Johnson, M.S.; Kulomaa, M.S.; Airenne, T.T.; Nordlund, H.R. Structural and functional characteristics of xenavidin, the first frog avidin from Xenopus tropicalis. *BMC Struct. Biol.* **2009**, *9*, 1–13. [CrossRef] [PubMed]
36. Chaiet, L.; Wolf, F.J. The properties of streptavidin, a biotin-binding protein produced by Streptomycetes. *Arch. Biochem. Biophys.* **1964**, *106*, 1–5. [CrossRef]
37. Helppolainen, S.H.; Nurminen, K.P.; Määttä, J.A.; Halling, K.K.; Slotte, J.P.; Huhtala, T.; Liimatainen, T.; Ylä-Herttuala, S.; Airenne, K.J.; Närvänen, A. Rhizavidin from Rhizobium etli: The first natural dimer in the avidin protein family. *Biochem. J.* **2007**, *405*, 397–405. [CrossRef] [PubMed]
38. Takakura, Y.; Tsunashima, M.; Suzuki, J.; Usami, S.; Kakuta, Y.; Okino, N.; Ito, M.; Yamamoto, T. Tamavidins–novel avidin-like biotin-binding proteins from the Tamogitake mushroom. *FEBS J.* **2009**, *276*, 1383–1397. [CrossRef] [PubMed]
39. Takakura, Y.; Sofuku, K.; Tsunashima, M.; Kuwata, S. Lentiavidins: Novel avidin-like proteins with low isoelectric points from shiitake mushroom (*Lentinula edodes*). *J. Biosci. Bioeng.* **2016**, *121*, 420–423. [CrossRef]

foods

Article

Design, Synthesis, and Characterization of Tracers and Development of a Fluorescence Polarization Immunoassay for Rapid Screening of 4,4′-Dinitrocarbanilide in Chicken Muscle

Qidi Zhang [1,†], Ming Zou [1,†], Wanyu Wang [1], Jinyan Li [1] and Xiao Liang [1,2,*]

1 College of Veterinary Medicine, Qingdao Agricultural University, No. 700 Changcheng Road, Qingdao 266109, China; zqdcau@qau.edu.cn (Q.Z.); mzou@qau.edu.cn (M.Z.); 20192113675@stu.qau.edu.cn (W.W.); 13730914323@163.com (J.L.)
2 Basic Medical College, Qingdao University, No. 308 Ningxia Road, Qingdao 266071, China
* Correspondence: 201401019@qau.edu.cn; Tel.: +86-139-64867-357
† These authors contributed equally to this work.

Abstract: The compound, 4,4′-dinitrocarbanilide (DNC), is the marker residue of concern in edible tissues of broilers fed with diets containing anticoccidial nicarbazin (NIC). In this study, 25 fluorescein-labeled DNC derivatives (tracers) are synthesized and characterized to develop a rapid fluorescence polarization immunoassay (FPIA) for the detection of DNC in chickens using DNC monoclonal antibodies (mAbs). The effect of the tracer structure on the sensitivity of the FPIA is investigated. Our results show that after optimization, the half maximal inhibitory concentrations (IC_{50}) and limit of detection (LOD) of the FPIA in the buffer are 28.3 and 5.7 ng mL^{-1}, respectively. No significant cross-reactivity (CR < 0.89%) with 15 DNC analogues is observed. The developed FPIA is validated for DNC detection in spiked chicken homogenates, and recoveries ranged from 74.2 to 85.8%, with coefficients of variation <8.6%. Moreover, the total time needed for the detection procedure of the FPIA, including sample pretreatment, is <40 min, which has not been achieved in any other immunoassays for DNC from literature. Our results demonstrate that the FPIA developed here is a simple, sensitive, specific, and reproducible screening method for DNC residues in chickens.

Keywords: DNC residue; FPIA; tracers; chicken muscle

Citation: Zhang, Q.; Zou, M.; Wang, W.; Li, J.; Liang, X. Design, Synthesis, and Characterization of Tracers and Development of a Fluorescence Polarization Immunoassay for Rapid Screening of 4,4′-Dinitrocarbanilide in Chicken Muscle. *Foods* **2021**, *10*, 1822. https://doi.org/10.3390/foods10081822

Academic Editor: Cristina A. Fente

Received: 7 July 2021
Accepted: 2 August 2021
Published: 6 August 2021

1. Introduction

Coccidiosis refers to the disease caused by protozoans of the genus *Eimeria* resulting in a wide range of injuries in the intestinal tracts of poultry [1,2]. Intestinal invasion by these protozoans disrupts feeding, digestive processes, and nutrient absorption and results in dehydration, blood loss, and increased susceptibility to other etiological agents. Together, these effects can result in significant economic losses in the poultry industry, as well as considerable cause distress to the animals [3].

Numerous types of coccidiostats have been developed, and nicarbazin (NIC) was the first such agent found to give satisfactory control of coccidiosis in chicken production facilities and has been in general use for this purpose since the 1960s. The most common form of NIC is an equimolar mixture of 4,4′-dinitrocarbanilide (DNC) and 2-hydroxy-4, 6-dimethyl pyrimidine (HDP) (Figure S1). Despite emerging drug resistance in protozoan parasite populations, NIC has maintained its effectiveness against all species of *Eimeria* [4]. However, NIC use as a coccidiostats is limited to the initial growth phases of chickens, since it causes adverse heat stress effects when administered to older birds and often results in their mortality. Moreover, long term human exposure to low levels of NIC can result in chronic toxicity [5]. In addition, NIC administered via feed in chickens results in the persistence of the DNC residues in edible tissues [6]. Hence, DNC is the marker residue used for the detection of NIC in edible chicken tissues. The maximum residue limits (MRL) of DNC in food matrices has been established by Food and Agriculture Organization (FAO), as well as by New Zealand for chicken muscle tissues at 200 μg kg^{-1} [7]. The MRL has been set at 200 μg kg^{-1} in China [8]. Japan has established an MRL of DNC at 20 μg Kg^{-1} for aquatic products.

The current analytic methods for the determination of DNC in food matrices utilize high performance liquid chromatography (HPLC) and HPLC-tandem mass spectrometry (MS/MS) [9–13]. These methods are accurate and sensitive, but require well-equipped laboratories, high capital expenditures, highly trained personnel, and generally involve time consuming sample preparation steps. Therefore, a rapid and efficient alternative detection method for screening large numbers of samples would streamline DNC screening efforts.

Immunoassay techniques are effective and economical alternatives to instrumental methods for DNC. The enzyme-linked immunosorbent assay (ELISA) is the most frequent choice [1,14,15], but it is a heterogeneous solid phase method and the time required to complete the assay is often >2 h, because of the need to separate the unbound probe in solution before the bound probe can be quantified [1]. In contrast, fluorescence polarization immunoassay (FPIA) is a competitive homogeneous assay that is based on differences in fluorescence polarization (FP) of the fluorescently-labeled analyte in the antibody bound and non-bound fractions (Figure S2). These reactions can reach equilibrium in minutes or even seconds, and no separation or washing steps are required, which makes it ideal for high-throughput screening of large numbers of samples [16,17]. These advantages of the FPIA have resulted in their widespread use in high-throughput screening of chemical contaminants, such as veterinary drugs [16–19], mycotoxins, pesticides, and other environmental contaminants in foods, feed, and environmental samples [20–22]. However, the application of FPIA to the detection of DNC has not been reported.

The top priorities in developing high-throughput screening methods are to simplify the assay and shorten analysis time, while maximizing the sensitivity of the methods [19]. With this aim, we developed a novel FPIA method for the screening of DNC in the present work. We investigated the effects of different tracers and physicochemical factors on the performance of the FPIA. The optimized immunoassay was compared with other published detection methods, and the newly developed method provided a simple, rapid, reproducible, and highly sensitive detection of DNC. Analysis of DNC in chicken samples required <40 min to complete using the FPIA.

2. Materials and Methods

2.1. Reagents and Equipments

NIC, 4-Nitroaniline, 2-Nitroaniline, 3-Nitroaniline, N-(4-Nitrophenyl) propionamide, H-Val-Pna HCl, L-Argininep-Nitroanilide Dihydrochloride, 4-Nitrophenethylamine hydrochloride, N-Methyl-4-nitrophenethylamine hydrochloride, H-Ala-Pna HCl, N,N-Dimethyl-4-Nitroaniline, H-Glu-Pna, Halofuginone, Toltrazuril, 1, 3-Diphenylguanidine, Ronidazole, and dinitolmide were obtained from Sigma-Aldrich (St. Louis, MO, USA). N, N-Dimethylformamide (DMF), 1-ethyl-3-(3-dimethylaminopropy) carbodiimide (EDC), N-hydroxysuccinimide (NHS), fluorescein isothiocyanate, ethylenediamine, butanediamine, hexamethylenediamine, 5-Aminofluorescein (5-AF), and 6-Aminofluorescein (6-AF) were obtained from Aladdin (Shanghai, China). Nonbinding-surface black microplates were purchased from Corning Life Sciences (New York, NY, USA). Normal solvents and salts were of analytical reagent grade and were supplied by Beijing Reagent Corporation (Beijing, China). The haptens DNC-1, DNC-2, DNC-3, DNC-4, and DNC-5 and four mAbs (4E1, 4B8, 2A12, and 3B4) were acquired from China Agricultural University [23]. Borate buffer (0.05 M, pH 8.0) was used as the working buffer for all FPIA experiments. The standard solution of DNC (2 mg mL^{-1}) was prepared by dissolving 4 mg of the DNC standard in 2 mL of dimethyl sulfoxide, and stored at −20 °C until use.

A SpectraMax M5 microplate reader from Molecular Devices (Downingtown, PA, USA) was used to measure FP. Black microplates (96-well) with a non-binding surface for FPIA were purchased from Corning Life Sciences (New York, NY, USA). Water was purified using a Milli-Q system from Millipore Inc. (Bedford, MA, USA). A ultraviolet analyzer was obtained from Tianjin Huike Instrument Equipment Co., Ltd. (Tianjin, China).

2.2. Preparation of FITC-DNC Tracers

Thiocarbamoyl ethylenediamine fluorescein (EDF), thiocarbamoyl butane diamine fluorescein (BDF), and thiocarbamoyl hexane fluorescein (HDF) were described previously [16]. Synthesis of the tracers used in the current study has been described elsewhere [24]. Briefly, 10 mg of hapten (DNC-1, DNC-2, DNC-3, DNC-4, or DNC-5) dissolved in 500 μL DMF was mixed with 30 mg NHS and 40 mg of EDC. After stirring at room temperature overnight, the reaction mixture was centrifuged to remove the precipitate at 5000× *g* for 10 min. Then 10 mg EDF was added to the supernatant and stirred overnight at room temperature. An aliquot (50 μL) of the mixture was purified using thin layer chromatography (TLC) with methanol/ trichloromethane (1:6, *v/v*) as eluent. The major TLC bands possessed a retardation factor (R_f) of 0.70, and they were collected and stored in methanol at 4 °C in the dark (Figure S3). In total, 25 tracers (Figure 1) for DNC were generated, and those bound to the specific mAbs were chosen for further study (see below).

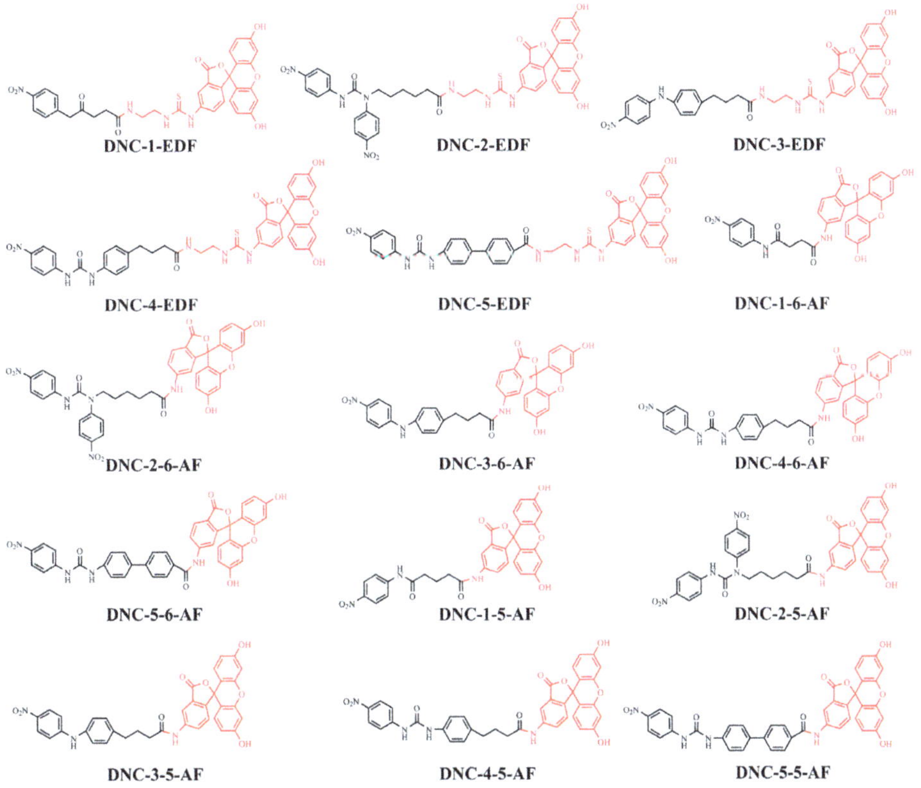

Figure 1. *Cont.*

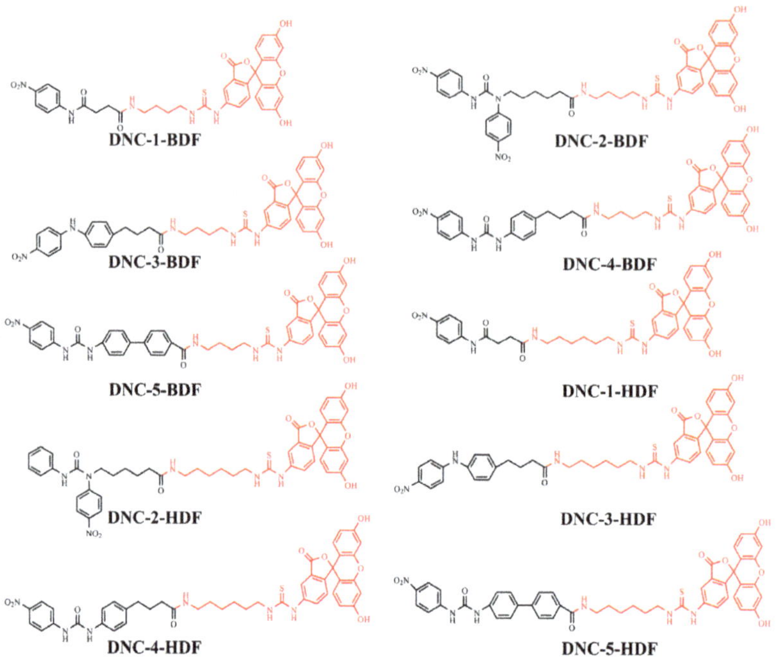

Figure 1. Structures of 25 different tracers of DNC.

2.3. FPIA Procedure

The tracer solutions were diluted with borate buffer to working concentrations with FP values 10 times that of the borate buffer background. The FPIA approach was described, as follows—70 μL of each tracer solution and 70 μL DNC standard solution or borate buffer were added to 70 μL antibody in a microplate well. After incubation for 20 min in the dark at room temperature, the FP value of the mixture was measured at λ_{ex} 485 nm and λ_{em} 530 nm (emission cutoff = 515 nm), respectively.

The antibody-tracer pair used in the final assay was the DNC-4-BDF and mAb 3B4 combination. The concentration of DNC-4-BDF was 200 RFU, and the dilution of mAb 3B4 was 1/300. The incubation time was 20 min at room temperature. The physicochemical conditions of the reaction buffer were pH 8 and NaCl concentration at 0 mM, respectively.

2.4. Curve Fitting and Statistical Analysis

A sigmoidal curve was used to fit FPIA data via OriginPro 8.0 (Origin Lab, Northampton, MA, USA) for construction of the standard curves. A four-parameter logistic equation was used to fit the immunoassay data as follows:

$$Y = (A - D)/[1 + (X/C)^{B}] + D \quad (1)$$

where A and D represent the maximum and minimum values, respectively; B is the slope factor; C is the concentration corresponding to 50% specific binding (IC_{50}), and X is the calibration concentration [25]. The limit of detection (LOD) was the concentration of the standard causing 10% inhibition of tracer binding (IC_{10}), and the working range corresponded to concentrations of the standard from IC_{20} to IC_{80} on the calibration curve. The IC_{50} and LOD values were used to evaluate the properties of the FPIA.

The specificity of the immunoassays was evaluated by determining the cross-reactivity (CR) with DNC analogs under the optimized conditions. CR was calculated with the following equation:

$$CR\ (\%) = (IC_{50}\ of\ DNC/IC_{50}\ of\ DNC\ analog) \times 100 \quad (2)$$

2.5. FPIA Development and Optimization

The standard test procedure is outlined in Section 2.3 (see above). In brief, respective antibody tracer pairs were selected using the detection window according to the equation $\delta mP = mP_{bound} - mP_{free}$ where IC_{50} values for the FPIA and the $IC_{50}/\delta mP$ ratio were used as the main parameters for the selection of optimum antibody tracer pairs. Tracer solutions were diluted with borate buffer to working concentrations that possessed fluorescence intensity (FI) values 10× above background. Specifically, diluted antibody and tracer that possessed a suitable δmP (≥90 mP) was evaluated, and the IC_{50} and $IC_{50}/\delta mP$ values were determined and adjusted to minimize the IC_{50} and $IC_{50}/\delta mP$ values and were calculated according to standard curves for DNC (see above). The FPIA was optimized by determining the influences of antibody dilution, reaction time, pH, salt concentration, and organic solvent levels on assay characteristics. The δmP and IC_{50} values were used as the primary criteria to evaluate FPIA performance.

2.6. Chicken Sample Analysis for FPIA

Samples of chicken muscle (2 g) were homogenized and then extracted with 2 mL methanol at room temperature. The mixtures were vortexed vigorously for 10 min and centrifuged at 10,000 rpm for 10 min at 4 °C. The supernatant was diluted 5-fold with assay buffer prior to analysis. For recovery experiments, blank chicken matrix samples obtained from WDWK Biotech (Beijing, China) were fortified with NIC at 50, 100, and 150 $\mu g\ kg^{-1}$, and 5 replicates were analyzed at each concentration using the optimized FPIA.

3. Results and Discussion

3.1. Synthesis and Characterization of FITC-DNC Tracers

The assay tracers EDF, BDF, and HDF possessed molecular ion peaks (m/z) of 450, 478.03, and 506.06, respectively, demonstrating that the fluorescein conjugates were synthesized successfully (Figure S4). These tracers play key roles in the FPIA, since both the hapten type and the bridge length between the hapten and the FITC can markedly influence antibody recognition. Therefore, we examined five structurally different haptens and five fluorescein molecules containing different carbon bridges and synthesized 25 tracers to evaluate FPIA performance. The high molecular polarities of the haptens and fluorescein resulted in high R_f values on TLC when methanol/trichloromethane (1:6, *v/v*) was used as the developing solvent. After the separation and purification of the tracers with TLC, they were characterized using FPIA. There were nine tracers that displayed obvious binding to the specific mAbs (δmP > 50 mP) and possessed significant immunochemical activity indicative of successful tracer conjugation (Figure 2). Within this group, DNC-1-EDF and DNC-5-EDF possessed the lowest δmP values (54 and 57 mP, respectively), indicating that they were not suitable for developing a sensitive FPIA. The remaining seven tracers (DNC-3-EDF, DNC-4-EDF, DNC-3-BDF, DNC-4-BDF, DNC-1-HDF, DNC-3-HDF, and DNC-4-HDF) and four MAbs (4E1, 4B8, 2A12 and 3B4) were used to construct antibody dilution curves (Figure 3A–D). Satisfactory binding (δmP = 94~236 mP) was observed for all these tracers when combained with each of the 4 mAbs. Thus, these seven tracers were selected for further study.

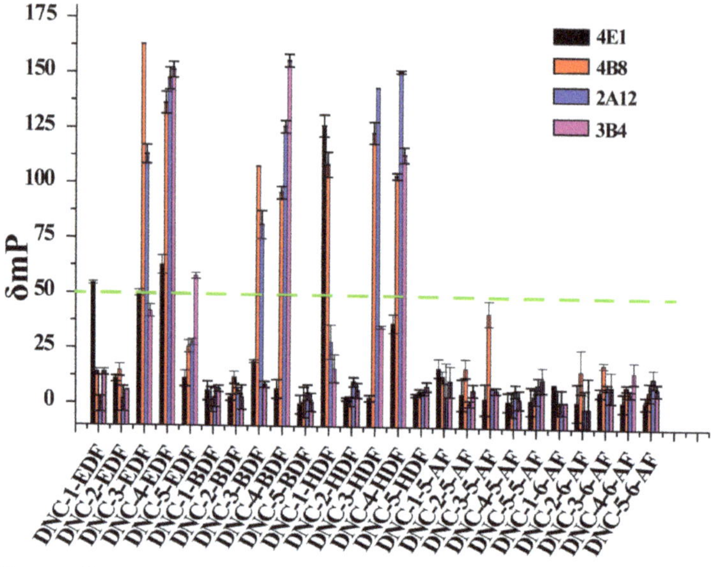

Figure 2. Characterization of the fluorescein-labeled DNC conjugates.

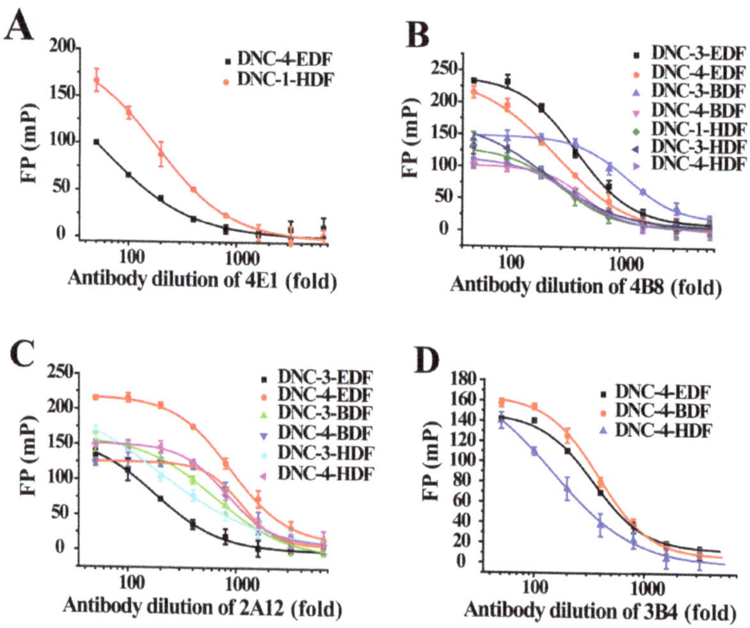

Figure 3. Antibody dilution curves of tracers. (**A**) 4E1 with DNC-4-EDF and DNC-1-HDF. (**B**) 4B8 with DNC-3-EDF, DNC-4-EDF, DNC-3-BDF, DNC-4-BDF, DNC-1-HDF, DNC-3-HDF, and DNC-4-HDF. (**C**) 2A12 with DNC-3-EDF, DNC-4-EDF, DNC-3-BDF, DNC-4-BDF, DNC-3-HDF, and DNC-4-HDF. (**D**) 3B4 with DNC-4-EDF, DNC-4-BDF, and DNC-4-HDF.

3.2. *FPIA Development and Optimization*

3.2.1. Selection of Antibody—Tracer Pairs

The combination of tracer and antibody has significant impacts on the sensitivity and specificity of an FPIA [26]. We evaluated the best tracer and antibody pair for use in the assay by constructing DNC standard curves. The optimum of mAb dilution was obtained from antibody dilution curves, and a δmP of 100 mP was set as the target detection range. In particular, the DNC-4-BDF and mAb 3B4 combination provided the lowest IC_{50} (66 ng mL^{-1}) and IC_{50}/δmP ratio (0.61) along with a broad detection window (δmP = 107 mP) (Figure 4A). Moreover, the Z′ factor of 0.89 represented a good separation for the distributions and indicated a robust FPIA. This combination was, therefore, selected to develop an FPIA for the detection of DNC (Table 1).

Table 1. Analytical parameters of the standard curves were obtained using four anti-NIC mAbs with seven DNC tracers.

Tracers	mAbs	Dilution Fold	IC_{50} (ng mL^{-1})	δmP	IC_{50}/δmP	Z′
DNC-3-EDF	2A12	100	1890	94	20.11	0.93
	4B8	400	26680	83.15	320.87	0.90
DNC-4-EDF	2A12	1800	5160	91.58	56.34	0.88
	4B8	500	263	96	2.74	0.87
	3B4	200	223	82.1	2.72	0.82
	4E1	50	7240	99.92	72.46	0.98
DNC-3-BDF	2A12	100	1520	98.72	15.39	0.92
	4B8	200	11360	71.45	158.99	0.94
DNC-4-BDF	2A12	400	1814	82.07	22.10	0.87
	4B8	100	3370	99	34.04	0.94
	3B4	200	66	107.76	0.61	0.89
DNC-1-HDF	4B8	200	91	89.28	1.02	0.84
	4E1	200	11600	75.65	153.33	0.81
DNC-3-HDF	2A12	200	290	133.07	2.18	0.91
	4B8	200	9960	68.85	144.66	0.87
DNC-4-HDF	2A12	700	4130	104.6	39.48	0.97
	4B8	200	4020	117.15	34.31	0.96
	3B4	100	2623	132.55	19.78	0.93

The use of heterologous tracers for FPIA significantly enhances its sensitivity [16,27], and we found similar results in our study. For example, mAb 4B8 was prepared using DNC-3-KLH as immunogen and was paired with the seven tracers [23]. The IC_{50} values (ng mL^{-1}) for the FPIA with these structurally heterologous tracers were DNC-4-EDF (263), DNC-4-BDF (3370), DNC-1-HDF (91), DNC-4-HDF (4020). These values were generally lower than those for the FPIA generated with homologous tracers, such as DNC-3-EDF (26680), DNC-3-BDF (11360), and DNC-3-HDF (9960) (Figure S5 and Table 1). When the remaining three mAbs (i.e., 2A12, 3B4, and 4E1) were examined in a similar way using DNC-4-BSA, DNC-5-BSA, and DNC-1-BSA as immunogens and then paired with the seven tracers, similar results were obtained because heterologous tracers could be more easily replaced by competitors (Table 1).

The effects of the length of the linker chain between a given hapten and fluorescein on FPIA sensitivity were also investigated in this study. The primary differences were the linker structure or length, as well as the orientation of the attached fluorophore. The IC_{50} values (ng mL^{-1}) for the FPIA increased when mAb 4B8 was paired with DNC-4-EDF (263), DNC-4-BDF (3370), and DNC-4-HDF (4020). The lowest IC_{50} was obtained with DNC-4-BDF when mAb 3B4 or 2A12 were paired with DNC-4-based tracers. The DNC-3-HDF tracer possessed the longest bridge and displayed the lowest IC_{50} when combined with mAb 4B8 or 2A12 (Table 1). Previous studies have indicated that sensitivity was optimal with long linkers between a hapten and fluorescein [19,28,29]. In contrast, FPIA based on short bridge tracers resulted in greater assay sensitivity [30,31]. Thus, both the structural features of the tracer hapten itself and the structure and length of the bridge between a hapten and fluorescein label markedly influence the recognition of the tracer by antibody, and it is necessary to select the optimal combination empirically.

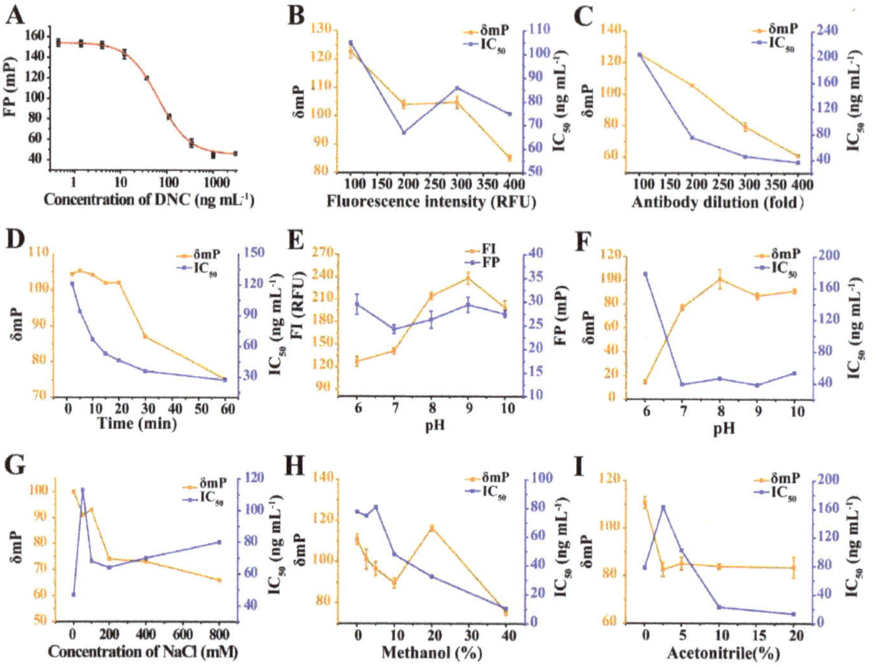

Figure 4. (**A**) Standard curves for DNC using DNC-4-BDF and mAb 3B4 combination. (**B**) Optimization of fluorescence intensity. (**C**) Optimization of antibody dilution. (**D**) Optimization of incubation time. (**E**) The effects of pH on FI and FP values. (**F**) Optimization of pH. (**G**) Optimization of NaCl concentration. (**H**) Effects of methanol concentration on assay performance. (**I**) Effects of acetonitrile concentration on assay performance.

3.2.2. Optimization of Tracer and mAb Concentrations

A competitive FPIA was used to optimize antibody and tracer concentrations. For instance, a low tracer concentration would yield higher sensitivity, but would also reduce the precision of the FP signal. In our assays, the FI values for DNC-4-BDF at levels of 5, 10, 15, and 20-fold greater than background (FI, 20 relative fluorescence units (RFU)) were examined. The optimal working concentration was set at an FI value of 200 RFU that was 10-fold higher than background as the lowest IC_{50} value, and an appropriate δmP value (104 mP) was observed (Figure 4B). The working concentration of mAb 3B4 was optimized at a fixed amount of tracer according to the standard curves for DNC. Both the IC_{50} and δmP values increased as mAb 3B4 dilution increased, and the lowest IC_{50} value was observed with a dilution of 1/300. This δmP value (<80 mP) was too low for FPIA, and therefore, a 1/200 for mAb 3B4 was used because it retained a lower IC_{50} value, and an acceptable δmP value of 105 mP (Figure 4C).

3.2.3. Optimization of Incubation Time

The incubation time for the assay must be chosen until equilibrium is established in the competition between an analyte and tracer [20]. We examined incubation times ranging from 2 to 60 min. It was found that the IC_{50} values decreased as incubation time increased from 2 min to 20 min and then plateaued at >20 min. The δmP value was almost constant with times <20 min and then decreased significantly at >20 min (Figure 4D). These results revealed that equilibrium was achieved at 20 min after the mixing of antibody and tracer; thus, 20 min was selected as the optimum incubation time.

3.2.4. Optimization of Physicochemical Conditions

The effects of pH, ionic strength, and organic solvent tolerance were also assessed for the FPIA by comparing IC_{50} and δmP values under various reaction conditions. The dyes used in this assay were pH-sensitive materials, and their FI values increased in the pH range of 6 to 8 and then decreased at pH >8, although the FP values were not affected significantly by pH (Figure 4E,F). These results indicated that pH 6 was not suitable for the assay system, due to insufficient binding (δmP = 14). This implied that the developed FPIA could not tolerate acidic conditions. The IC_{50} values changed little at pH values ranging from 7 to 10, and the assay performed optimally at pH 8 with high sensitivity and maximal δmP (>100 mP). NaCl concentrations between 0 to 800 mM and negative effects on the IC_{50} and δmP were observed for NaCl concentrations from 0 to 800 mM (Figure 4G). High ionic strength most likely resulted in the disruption of antibody-antigen interactions. Therefore, NaCl was not included in the assay buffer. Methanol and acetonitrile added to the test system led to IC_{50} and δmP decreased in direct proportion to their concentrations. The presence of 20% methanol resulted in a significant increase of δmP (116 mP), and this solvent was generally better tolerated than acetonitrile. Acetonitrile concentrations exceeding 20% (final concentration) were not tolerated, although 20% acetonitrile was still tolerated (Figure 4H,I). If needed, the assay could be used at methanol concentrations of up to 20%.

3.3. Sensitivity and Cross-Reactivity of FPIA

Under the optimal conditions, the sensitivity and specificity of the FPIA were calculated using DNC and 15 DNC analogs. The IC_{50} value, the LOD defined as IC_{10} from the standard curve and linear range (IC_{20}~IC_{80}) were 28.39, 5.70, 10.31~78.17 ng mL^{-1}, respectively (Figure 5A). The CR values for the developed FPIA were <0.89% for all 15 NIC analogs (Table 2). These results indicated that this FPIA was highly sensitive and specific for DNC.

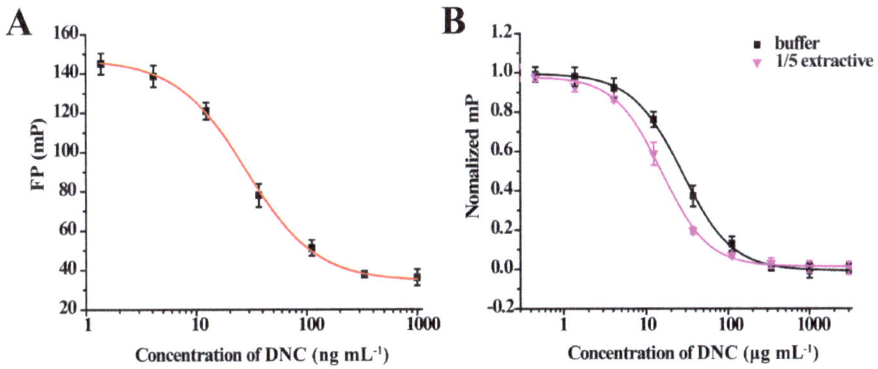

Figure 5. (**A**) Standard curves for NIC using the FPIA in assay buffer under optimized conditions. (**B**) Comparison of FPIA curves obtained from standards prepared in assay buffer, and 5-fold diluted chicken extracts.

Table 2. IC_{50} values and Cross-reactivity of DNC and 15 structurally related analogs for the FPIA.

Analogues	Structure	IC_{50} (ng mL^{-1})	CR (%)
DNC	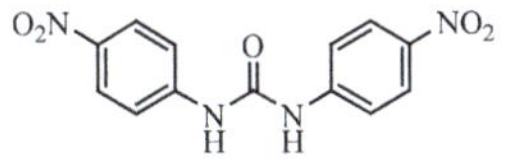	28.39	100

Table 2. *Cont.*

Analogues	Structure	IC_{50} (ng mL^{-1})	CR (%)
2-Nitroaniline	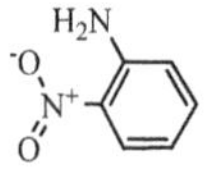	>30,000	<0.1
3-Nitroaniline	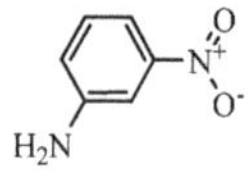	>30,000	<0.1
N-(4-Nitrophenyl) propionamide	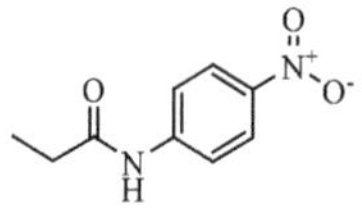	>30,000	<0.1
H-Val-Pna HCl	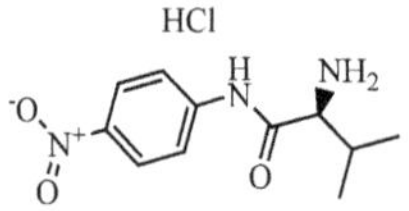	>30,000	<0.1
L-Arginine P-Nitroanilide Dihydrochl Oride	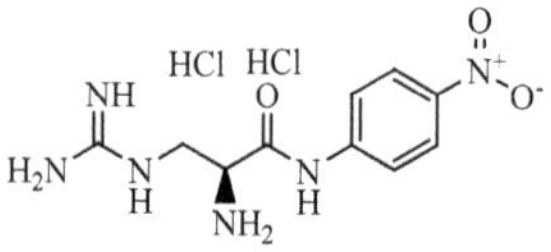	>30,000	<0.1
4-Nitrophenethylamine hydrochloride	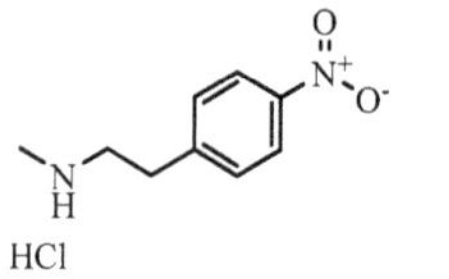	>30,000	<0.01
N-Methyl-4-nitrophenethylamine hydrochloride	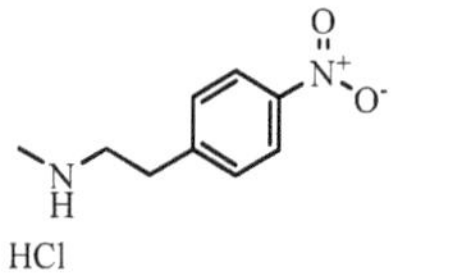	>30,000	<0.1
H-Ala-Pna HCl	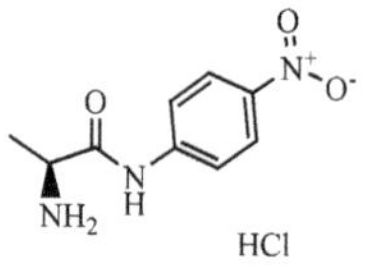	>30,000	<0.1
N, N-Dimethyl-4-Nitroaniline	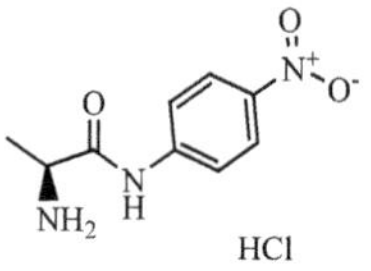	26,435	0.11

Table 2. *Cont.*

Analogues	Structure	IC_{50} (ng mL^{-1})	CR (%)
H-Glu-Pna	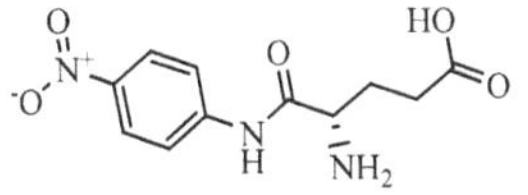	>30,000	<0.1
Halofuginone	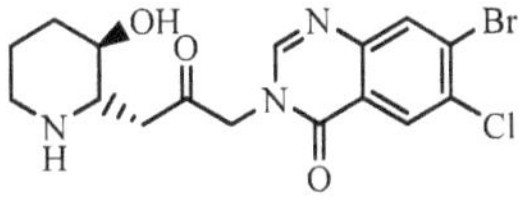	>30,000	<0.1
Toltrazuril	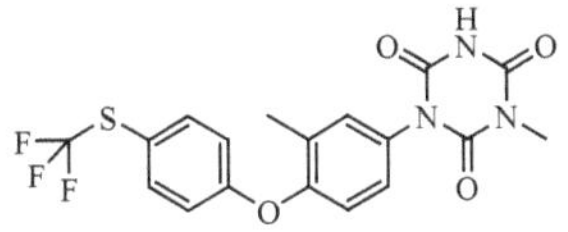	3190	0.89
1,3-Diphenylguanidine	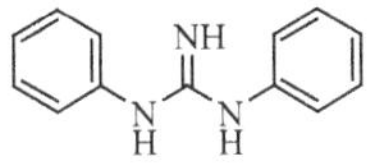	>30,000	<0.1
Ronidazole	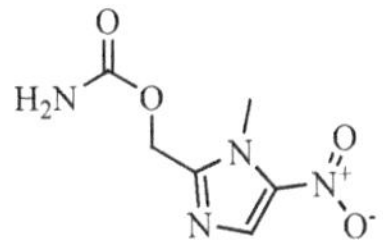	>30,000	<0.1
Dinitolmide	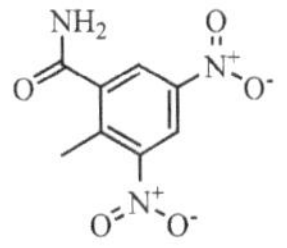	>30,000	<0.1

3.4. Chicken Sample Analysis for FPIA

To further evaluate the utility of the FPIA, we determined recoveries of spiked tissue matrices. Calibration curves prepared in buffer, and diluted chicken extract were superimposed with normalization of the FP values indicating that the matrix interference was eliminated using 5-fold dilutions in methanol (Figure 5B). The mean recovery values for DNC at 50, 100, and 150 μg kg^{-1} ranged from 74.23 to 85.80% with CVs <8.64%. The LOD was 24.21 μg kg^{-1} and was sensitive enough to meet the detection requirements of MRL for DNC in chicken tissues set by the EU, the USA, and China. The working range of the FPIA was 31.15 to 188.35 μg kg^{-1} (Table 3). The sensitivity of the FPIA was lower than the ELISA developed using the same mAb (3B4) [23], but the FPIA, a homogeneous method, requires a much shorter time (less than 40 min) for the detection of DNC in chicken muscle. This is an essential characteristic needed for a rapid screening method. Compared with the required time of other immunoassays for DNC in animal-derived food or feeds, including ELISA (>2 h) [15], surface plasmon resonance (SPR) biosensor screening (>49 min) [32], time-resolved fluoroimmunoassays (TR-FIA) (>2.5 h) [33], and flow cytometry-based immunoassay (>2 h) [34], the shortest time was required for the new developed FPIA for DNC in chicken muscle.

Table 3. Recovery studies from chicken muscle matrices using FPIA.

Sample	Spiked (μg kg^{-1})	Intra-Assay (n = 5)		Inter-Assay (n = 5)	
		Recovery (%)	CV (%)	Recovery (%)	CV (%)
Chicken	50	84.41	4.39	80.45	4.81
	100	82.6	5.12	85.80	4.00
	150	74.23	8.64	76.95	6.41

4. Conclusions

In summary, an FPIA for DNC was established with favorable sensitivity, specificity, cost, time, and reliability for the first time. The sensitivity of the developed FPIA was significantly improved by optimizing the selection of tracers, tracer-antibody pairs, and physical and chemical reaction conditions. Furthermore, the reliability and robustness of the assay were successfully demonstrated for analysis of DNC in chicken muscle matrices. In addition, the sample pretreatment was simple for the developed FPIA. The total analysis time, including sample pretreatment, was less than 40 min, which has not yet been achieved in other immunoassays for DNC residues.

Supplementary Materials: The following are available online at https://www.mdpi.com/article/10.3390/foods10081822/s1, Figure S1. The chemical structures of DNC and HDP, Figure S2. Schematic illustrations of the FPIA, Figure S3. The results of TLC for 25 tracers. (The red clip represents the target band that binds to the antibody), Figure S4. (A) Mass spectrum of EDF with molecular ion peak $[M + 2H]^{2+}$ (m/z) of 450.00; (B) Mass spectrum of BDF with molecular ion peak $[M + 2H]^{2+}$ (m/z) of 478.03; (C) Mass spectrum of HDF with molecular ion peak $[M + 2H]^{2+}$ (m/z) of 506.06, Figure S5. Comparison of the effect of homologous and heterologous tracers on sensitivity.

Author Contributions: Conceptualization, X.L.; methodology, W.W.; software, Q.Z. and M.Z.; validation, W.W.; formal analysis, W.W.; investigation, J.L.; resources, J.L.; data curation, Q.Z. and M.Z.; writing—original draft preparation, Q.Z. and M.Z.; writing—review and editing, X.L.; visualization, Q.Z. and M.Z.; supervision, X.L.; project administration, X.L.; funding acquisition, X.L. All authors have read and agreed to the published version of the manuscript.

Funding: This research was funded by Natural Science Foundation of Shandong Province (ZR2020MC187), Postdoctoral Innovation Project of Shandong Province (202002027), Postdoctoral Fund of Qing-dao, High Level Talent Fund of Qingdao Agricultural University (663/1115026), the Scientific and Technological Projects of Qingdao (19-6-1-94-nsh and 21-1-4-ny-10-nsh) and Natural Science Foundation of Shandong Province (ZR2020MC188).

Institutional Review Board Statement: Not applicable.

Informed Consent Statement: Not applicable.

Data Availability Statement: Not applicable.

Conflicts of Interest: The authors declare no conflict of interest.

Ethical Approval: This article does not contain any studies with human and animal subjects.

References

1. Beier, R.C.; Ripley, L.H.; Young, C.R.; Kaiser, C.M. Production, Characterization, and Cross-Reactivity Studies of Monoclonal Antibodies against the Coccidiostat Nicarbazin. *J. Agric. Food Chem.* **2001**, *49*, 4542–4552. [CrossRef]
2. Clarke, L.; Fodey, T.L.; Crooks, S.R.; Moloney, M.; O'Mahony, J.; Delahaut, P.; O'Kennedy, R.; Danaher, M. A review of coccidiostats and the analysis of their residues in meat and other food. *Meat Sci.* **2014**, *97*, 358–374. [CrossRef]
3. Protasiuk, E.; Olejnik, M.; Szprengier–Juszkiewicz, T.; Jedziniak, P.; Zmudzki, J. Determination of Nicarbazin in Animal Feed by High-Performance Liquid Chromatography with Interlaboratory Evaluation. *Anal. Lett.* **2015**, *48*, 2183–2194. [CrossRef]
4. Bacila, D.M.; Feddern, V.; Mafra, L.I.; Scheuermann, G.N.; Molognoni, L.; Daguer, H. Current research, regulation, risk, analytical methods and monitoring results for nicarbazin in chicken meat: A perspective review. *Food Res. Int.* **2017**, *99*, 31–40. [CrossRef]

5. Bacila, D.M.; Lazzarotto, M.; Hansel, F.A.; Scheuermann, G.N.; Feddern, V.; Cunha Junior, A.; Igarashi-Mafra, L. Thermal profile of 4,4′-dinitrocarbanilide determined by thermogravimetry–differential scanning calorimetry–mass spectrometry (TG–DSC–MS) and pyrolysis–gas chromatography–mass spectrometry (Py–GC–MS). *J. Therm. Anal. Calorim.* **2019**, *138*, 697–701. [CrossRef]
6. Tarbin, J.A.; Bygrave, J.; Bigwood, T.; Hardy, D.; Rose, M.; Sharman, M. The effect of cooking on veterinary drug residues in food: Nicarbazin (dinitrocarbanilide component). *Food Addit. Contam.* **2005**, *22*, 1126–1131. [CrossRef] [PubMed]
7. Capurro, E.; Danaher, M.; Anastasio, A.; Cortesi, M.L.; O'Keeffe, M. Efficient HPLC method for the determination of nicarbazin, as dinitrocarbanilide in broiler liver. *J. Chromatogr. B* **2005**, *822*, 154–159. [CrossRef]
8. Mortier, L.; Daeseleire, E.; Van Peteghem, C. Liquid chromatographic tandem mass spectrometric determination of five coccidiostats in poultry eggs and feed. *J. Chromatogr. B* **2005**, *820*, 261–270. [CrossRef] [PubMed]
9. Teglia, C.M.; Gonzalo, L.; Culzoni, M.J.; Goicoechea, H.C. Determination of six veterinary pharmaceuticals in egg by liquid chromatography: Chemometric optimization of a novel air assisted-dispersive liquid-liquid microextraction by solid floating organic drop. *Food Chem.* **2019**, *273*, 194–202. [CrossRef] [PubMed]
10. Barreto, F.; Ribeiro, C.; Hoff, R.B.; Costa, T.D. A simple and high-throughput method for determination and confirmation of 14 coccidiostats in poultry muscle and eggs using liquid chromatography–quadrupole linear ion trap-tandem mass spectrometry (HPLC–QqLIT-MS/MS): Validation according to European Union 2002/657/EC. *Talanta* **2017**, *168*, 43–51.
11. Bacila, D.; Cunha, A., Jr.; Gressler, V.; Scheuermann, G.; Coldebella, A.; Caron, L.; Mafra, L.; Feddern, V. Detection of p-Nitroaniline Released from Degradation of 4,4′-Dinitrocarbanilide in Chicken Breast during Thermal Processing. *J. Agric. Food Chem.* **2019**, *67*, 9002–9008. [CrossRef]
12. de Lima, A.L.; Barreto, F.; Rau, R.B.; da Silva, G.R.; Lara, L.J.C.; de Figueiredo, T.C.; de Assis, D.C.S.; de Vasconcelos Cançado, S. Determination of the residue levels of nicarbazin and combination nicarbazin-narasin in broiler chickens after oral administration. *PLoS ONE* **2017**, *12*, e0181755.
13. Moretti, S.; Fioroni, L.; Giusepponi, D.; Pettinacci, L.; Saluti, G.; Galarini, R. Development and validation of a multiresidue liquid chromatography/tandem mass spectrometry method for 11 coccidiostats in feed. *J. AOAC Int.* **2013**, *96*, 1245–1257. [CrossRef]
14. Connolly, L.; Fodey, T.L.; Crooks, S.R.H.; Delahaut, P.; Elliott, C.T. The production and characterisation of dinitrocarbanilide antibodies raised using antigen mimics. *J. Immunol. Methods* **2002**, *264*, 45–51. [CrossRef]
15. Huet, A.C.; Mortier, L.; Daeseleire, E.; Fodey, T.; Elliott, C.; Delahaut, P. Screening for the coccidiostats halofuginone and nicarbazin in egg and chicken muscle: Development of an ELISA. *Food Addit. Contam.* **2005**, *22*, 128–134. [CrossRef] [PubMed]
16. Mi, T.; Wang, Z.; Eremin, S.A.; Shen, J.; Zhang, S. Simultaneous Determination of Multiple (Fluoro) quinolone Antibiotics in Food Samples by a One-Step Fluorescence Polarization Immunoassay. *J. Agric. Food Chem.* **2013**, *61*, 9347–9355. [CrossRef] [PubMed]
17. Wang, Z.; Liang, X.; Wen, K.; Zhang, S.; Li, C.; Shen, J. A highly sensitive and class-specific fluorescence polarisation assay for sulphonamides based on dihydropteroate synthase. *Biosens. Bioelectron.* **2015**, *70*, 1–4. [CrossRef]
18. Gasilova, N.V.; Eremin, S.A. Determination of chloramphenicol in milk by a fluorescence polarization immunoassay. *J. Anal. Chem.* **2010**, *65*, 255–259. [CrossRef]
19. Dong, B.; Zhao, S.; Li, H.; Wen, K.; Ke, Y.; Shen, J.; Zhang, S.; Shi, W.; Wang, Z. Design, synthesis and characterization of tracers and development of a fluorescence polarization immunoassay for the rapid detection of ractopamine in pork. *Food Chem.* **2019**, *271*, 9–17. [CrossRef] [PubMed]
20. Ma, M.; Chen, M.; Feng, L.; You, H.; Yang, R.; Boroduleva, A.; Hua, X.; Eremin, S.A.; Wang, M. Fluorescence Polarization Immunoassay for Highly Efficient Detection of Imidaclothiz in Agricultural Samples. *Food Anal. Methods* **2016**, *9*, 2471–2478. [CrossRef]
21. Li, C.; Wen, K.; Mi, T.; Zhang, X.; Zhang, H.; Zhang, S.; Shen, J.; Wang, Z. A universal multi-wavelength fluorescence polarization immunoassay for multiplexed detection of mycotoxins in maize. *Biosens. Bioelectron.* **2016**, *79*, 258–265. [CrossRef]
22. Raysyan, A.; Moerer, R.; Coesfeld, B.; Eremin, S.A.; Schneider, R.J. Fluorescence polarization immunoassay for the determination of diclofenac in wastewater. *Anal. Bioanal. Chem.* **2021**, *413*, 999–1007. [CrossRef]
23. Qianqian Tang, Z.W. *Production of Monoclonal Antibodies and Development of an Enzyme-Linked Immunosobent Assay for Nicarbazin*; College of Veterinary Medicine, China Agricultural University: Beijing, China, 2018.
24. Mi, T.; Liang, X.; Ding, L.; Zhang, S.; Eremin, S.A.; Beier, R.C.; Shen, J.; Wang, Z. Development and optimization of a fluorescence polarization immunoassay for orbifloxacin in milk. *Anal. Methods* **2014**, *6*, 3849–3857. [CrossRef]
25. Li, H.; Ma, S.; Zhang, X.; Li, C.; Dong, B.; Mujtaba, M.G.; Wei, Y.; Liang, X.; Yu, X.; Wen, K.; et al. Generic Hapten Synthesis, Broad-Specificity Monoclonal Antibodies Preparation, and Ultrasensitive ELISA for Five Antibacterial Synergists in Chicken and Milk. *J. Agric. Food Chem.* **2018**, *66*, 11170–11179. [CrossRef] [PubMed]
26. Zhang, X.; Eremin, S.A.; Wen, K.; Yu, X.; Li, C.; Ke, Y.; Jiang, H.; Shen, J.; Wang, Z. Fluorescence Polarization Immunoassay Based on a New Monoclonal Antibody for the Detection of the Zearalenone Class of Mycotoxins in Maize. *J. Agric. Food Chem.* **2017**, *65*, 2240–2247. [CrossRef]
27. Wang, Z.; Zhang, S.; Ding, S.; Eremin, S.A.; Shen, J. Simultaneous determination of sulphamerazine, sulphamethazine and sulphadiazine in honey and chicken muscle by a new monoclonal antibody-based fluorescence polarisation immunoassay. *Food Addit. Contam. Part A Chem.* **2008**, *25*, 574–582. [CrossRef] [PubMed]
28. Yakovleva, J.; Zeravik, J.; Michura, I.; Formanovsky, A.; Fránek, M.; Eremin, S. Hapten Design and Development of Polarization Fluoroimmunoassay for Nonylphenol. *Int. J. Environ. Anal. Chem.* **2003**, *83*, 597–607. [CrossRef]

29. Krasnova, A.I.; Eremin, S.A.; Natangelo, M.; Tavazzi, S.; Benfenati, E. A polarization fluorescence immunoassay for the herbicide propanil. *Anal. Lett.* **2001**, *34*, 2285–2301. [CrossRef]
30. Chun, H.S.; Choi, E.H.; Chang, H.-J.; Choi, S.-W.; Eremin, S.A. A fluorescence polarization immunoassay for the detection of zearalenone in corn. *Anal. Chim. Acta* **2009**, *639*, 83–89. [CrossRef] [PubMed]
31. Wang, Q.; Haughey, S.A.; Sun, Y.-M.; Eremin, S.A.; Li, Z.-F.; Liu, H.; Xu, Z.-L.; Shen, Y.-D.; Lei, H.-T. Development of a fluorescence polarization immunoassay for the detection of melamine in milk and milk powder. *Anal. Bioanal. Chem.* **2011**, *399*, 2275–2284. [CrossRef] [PubMed]
32. McCarney, B.; Traynor, I.M.; Fodey, T.L.; Crooks, S.R.H.; Elliott, C.T. Surface plasmon resonance biosensor screening of poultry liver and eggs for nicarbazin residues. *Anal. Chim. Acta* **2003**, *483*, 165–169. [CrossRef]
33. Hagren, V.; Crooks, S.R.; Elliott, C.T.; Lövgren, T.; Tuomola, M. An all-in-one dry chemistry immunoassay for the screening of coccidiostat nicarbazin in poultry eggs and liver. *J. Agric. Food Chem.* **2004**, *52*, 2429–2433.
34. Campbell, K.; Fodey, T.; Flint, J.; Danks, C.; Danaher, M.; O'Keeffe, M.; Kennedy, D.G.; Elliott, C. Development and validation of a lateral flow device for the detection of nicarbazin contamination in poultry feeds. *J. Agric. Food Chem.* **2007**, *55*, 2497–2503. [CrossRef] [PubMed]

foods

MDPI

Article

Lateral Flow Immunochromatography Assay for Detection of Furosemide in Slimming Health Foods

Yingying Li, Haihuan Xie, Jin Wang, Xiangmei Li, Zhili Xiao, Zhenlin Xu, Hongtao Lei and Xing Shen *

Guangdong Provincial Key Laboratory of Food Quality and Safety, South China Agricultural University, Guangzhou 510642, China; wzlyy@stu.scau.edu.cn (Y.L.); xiehaihuan@stu.scau.edu.cn (H.X.); wangjin940810@stu.scau.edu.cn (J.W.); lixiangmei12@scau.edu.cn (X.L.); scauxzl@scau.edu.cn (Z.X.); jallent@163.com (Z.X.); hongtao@scau.edu.cn (H.L.)

* Correspondence: shenxing325@163.com; Tel.: +86-135-6043-2677

Abstract: In recent years, furosemide has been found to be abused in slimming health foods. There is an urgent need for a simpler, faster method for detecting furosemide in slimming health foods. In this study, a rapid, convenient and sensitive lateral flow immunochromatography (LFIA) based on Au nanoparticles (AuNPs) was established for the first time. Under optimal conditions, the qualitative limit of detection (LOD) of the AuNPs-based LFIA was 1.0~1.2 μg/g in slimming health foods with different substrates. AuNPs-LFIA could specifically detect furosemide within 12 min (including sample pretreatment) and be read by the naked eye. The developed AuNPs-LFIA showed high consistency with liquid chromatography with tandem mass spectrometry (LC-MS/MS), and no false positive or false negative results were found in spiked slimming health foods, proving that the AuNPs-LFIA should be accurate and reliable. The AuNPs-LFIA reported here provides a serviceable analytical tool for the on-site detection and rapid initial screening of furosemide for the first time.

Keywords: furosemide; Au nanoparticles; lateral flow immunochromatography; slimming health food

Citation: Li, Y.; Xie, H.; Wang, J.; Li, X.; Xiao, Z.; Xu, Z.; Lei, H.; Shen, X. Lateral Flow Immunochromatography Assay for Detection of Furosemide in Slimming Health Foods. *Foods* **2021**, *10*, 2041. https://doi.org/10.3390/foods10092041

Academic Editor: Thierry Noguer

Received: 24 July 2021
Accepted: 23 August 2021
Published: 30 August 2021

1. Introduction

Overweight and obesity are the main risk factors for many chronic diseases. According to a World Health Organization (WHO) report, 39% of the global population aged 18 and over were overweight in 2016, while 13% (male 11%, female 15%) were obese [1]. In order to lose weight quickly, people are increasingly turning to the consumption of slimming dietary supplements. To pursue economic benefits, some slimming dietary supplements have been illegally adulterated with synthetic drugs to obtain obvious short-term effects [2,3].

According to the current legislation of the European Union (EU), the USA and China, synthetic drugs are not allowed in dietary supplements due to their harmful side effects [4]. However, some unscrupulous traders continue to illegally add drugs to slimming dietary supplements to increase the weight loss effect for the purpose of promoting sales, especially diuretics, appetite suppressants, gastrointestinal lipase inhibitors, energy expenditure agents and laxatives. Diuretics are common adulterants in slimming health foods. They accelerate the excretion of water from the body, causing the illusion of weight loss. Consumers could purchase and take slimming health foods containing diuretics without knowing it. Overdosing these products can produce side effects such as fluid and electrolyte abnormalities as well as acid–base disturbances, which may cause severe arrhythmia and increase the risk of death from arrhythmia [5]. Furosemide is one of the most effective diuretic medications available. It acts directly on the kidneys to increase urine output and the urinary excretion of sodium [6]. Oral formulations of furosemide are commonly used to treat edema, congestive heart failure, renal failure and hypertension [7]. In recent years, it has often been found to be illegally added to weight loss health foods. In 2020, the Institute for Drug Control of Suzhou, China, tested 84 batches of slimming health foods, and the illegal addition of furosemide was detected in 13 batches of samples (positivity rate, 15.5%) [8]. Therefore, it is necessary to establish a detection method for the illegally added drug furosemide in health foods.

At present, various methods for detecting furosemide in slimming products have been reported, including capillary electrophoresis (CE) [9–11], ion migration spectrometry (IMS) [12], ion-pair chromatography (IPC) [13,14], liquid chromatography-tandem mass spectrometry (LC-MS/MS) [15], high performance liquid chromatography (HPLC) [16] and ultra-high-pressure liquid chromatography (UHPLC) [17,18]. All these methods rely on expensive equipment, which is difficult to operate and requires trained operators. Although their instrumental methods are accurate, they can not meet the requirements of rapid on-site inspection. The rapid detection of furosemide mainly includes electrochemical and immunoassay methods. Electrochemical analysis methods are mainly used to detect furosemide in urine and drugs, and have a good detection speed, detection sensitivity and detection throughput [6,19–21]; however, all of them lack simplicity and selectivity to the negatively charged furosemide. Immunoassay is a rapid analysis method that is currently widely used. By now, only two enzyme-linked immune sorbent assays (ELISAs) have been reported for detecting furosemide in horse plasma and milk [22–24]. However, ELISA also involves complex testing procedures and long incubation times, so it remains a laboratory-based platform unsuited to on-site detection. A simpler and faster on-site detection method is needed for monitoring the growing number of slimming products.

Lateral flow immunochromatography assay (LFIA), which is simple, rapid and low-cost, has been widely used in food safety, environmental monitoring and medical diagnosis in recent years [25–29]. AuNPs have many advantages as a mature labeling material, such as simple preparation, short labeling time, good stability and low cost [30,31]. Thus, they are favored by manufacturers and occupy more than 90% of the label market in LFIA [32]. In this paper, a convenient AuNPs-LFIA detecting furosemide with good sensitivity and specificity was developed for the first time and proved to be efficient for application in the detection of furosemide in slimming health foods.

2. Materials and Methods

2.1. Materials

Furosemide, goat anti-rabbit IgG (secondary antibody), bovine albumin (BSA), ovalbumin (OVA), *N*,*N*-dimethylformamide (DMF), 1-(3-fdimethylaminopropyl)-3-ethylcarbodiimide hydrochloride (EDC) and ProClin 300 were purchased from Sigma-Aldrich (St. Louis, MO, USA). Chloroauric acid, trisodium citrate and polyvinyl pyrrolidone (PVP) were purchased from Sinopharm Chemical Reagent Co., Ltd. (Shanghai, China). Anti-furosemide antibodies and coating antigens were prepared in our own laboratory. Other chemicals were purchased from Guangzhou Chemical Reagent Co., Ltd. (Guangzhou, China). All reagents were of analytical grade or higher purity. The nitrocellulose filter (NC) membrane (CN95) was obtained from Sartorius Stedim Biotech GmbH (Goettingen, Germany). The sample pad (blood filtration membranes) and the polyvinyl chloride (PVC) backing plate (SMA31-40) were purchased from Shanghai Liangxin Co., Ltd. (Shanghai, China).

2.2. Instruments

An FEI/Talos L120C transmission electron microscope (TEM) (Thermo Scientific, Waltham, USA) was used to observe the morphologies of nanoparticles. The zeta potential was measured by a Zetasizer Nano ZS90 (Malvern Panalytical, UK). The XYZ 3060 Dispensing Platform (BioDot, Irvine, CA, USA) was used to spray antigen and secondary antibodies onto the NC membrane. The strip cutter ZQ 2000 (Kinbio Tech, Shanghai, China) was used to cut test strips into suitable sizes. LC-MS/MS was carried out on an AB QTRAP4500 triple quadrupole mass spectrometer (SCIEX, Framingham, MA, USA).

2.3. Preparation of Coating Antigen

The coating antigen was obtained by furosemide coupled with cationized ovalbumin (cOVA). cOVA is obtained by modifying OVA with ethylenediamine. Furosemide contains a carboxyl group, which could be directly coupled with cOVA by the active ester method to produce a coating antigen. Furosemide (1 equiv.), N-Hydroxy succinimide (NHS) (1.5 equiv.) and 1-(3-fdimethylaminopropyl)-3-ethylcarbodiimide hydrochloride (EDC) (1.5 equiv.) were dissolved in 200 μL of *N,N*-dimethylformamide (DMF). The mixture was stirred at 4 °C for 6 h, and then centrifuged at 2500× *g* for 10 min. The supernatant was added dropwise to cOVA (20 mg) in 5 mL of PBS (phosphate-buffered saline, 0.01 M, pH 7.4). The conjugate mixture was stirred at 4 °C overnight and dialyzed against PBS (0.01 M, pH 7.4) for 3 days at 4 °C to remove the uncoupled free hapten and non-reacted reactants. The obtained conjugate was used as coating antigen.

2.4. Preparation of AuNPs

The AuNPs were produced by reducing $HAuCl_4$ with sodium citrate according to a previous method, which was modified as described below [33]. An amount of 8 mL of 1% chloroauric acid solution was quickly added into 200 mL of boiling ultrapure water under continuous stirring. When the solution boiled again, 9.25 mL of 1% trisodium citrate was added. The solution was then stirred and heated for another 10 min. After cooling, transmission electron microscopy and UV–visible absorption spectrometry were used to characterize the morphologies of AuNPs. The prepared AuNPs were stored at 4 °C for use.

2.5. Preparation of AuNPs–Abs Conjugated Probe

The AuNPs–Abs conjugated probe was prepared via electrostatic adsorption between AuNPs and antibodies (Figure 1a). The optimal labeling pH and the antibody amount were adjusted by checkerboard titration. A suitable amount of 0.2 M K_2CO_3 was added into the AuNPs solution to adjust the pH value. Anti-furosemide antibody dissolved in 100 μL of 0.01 M PB (phosphate buffer solution, pH 7.4) was quickly added into the pH-adjusted AuNPs solution. The mixture was reacted for 10 min at room temperature. Then, 20 μL of 20% BSA was added and incubated for 20 min to block excess binding sites on the AuNPs. After centrifuging at 10,000× *g* and 4 °C for 10 min, the supernatant was discarded, the bottom red precipitate was resuspended with 200 μL of resuspension buffer (0.005 M borate buffer solution, pH 8.0, containing 0.5% BSA, 5% trehalose for protecting antibody, 0.5% Tween-20 both for a better release AuNPs–Abs probe and to adjust the chromatography speed, 0.3% PVP as a steric stabilizer or capping agent to protect the AuNPs–Abs against agglomeration, and 0.03% ProClin 300 to prevent metamorphism), and finally stored at 4 °C for further use.

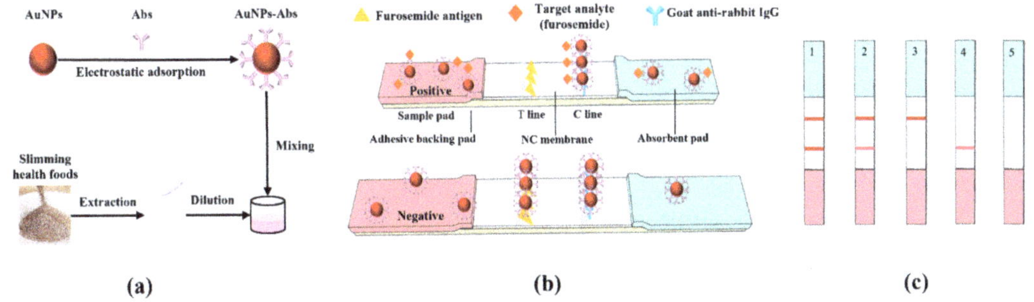

Figure 1. Schematic of the Au nanoparticles lateral flow immunochromatography (AuNPs-LFIA) for detecting furosemide in slimming coffee. (**a**) Preparation of the signal probe AuNPs–Abs and the sample treatment solution. (**b**) The structure and test procedure of the AuNPs-LFIA test strip. C line: control line (goat anti-rabbit immunoglobulin G, IgG) and T line: test line (furosemide coating antigen). (**c**) Schematic diagram of AuNPs-LFIA strip test results: 1, negative result; 2, weak positive result; 3, positive result; 4–5, invalid result.

To better reflect the performance of the AuNPs–Abs conjugated probe, a series of influencing parameters were optimized, including the pH value, the concentration and dilution buffer of antibody and antigen, and the resuspension buffer of AuNPs–Abs. The optimal conditions were selected according to the T line color intensity and sensitivity (inhibition rate, $(1\text{-}OD_{positive}/OD_{negative}) \times 100\%$).

2.6. Strip Assembly

The test strip of the LFIA was composed of an NC membrane, a sample pad, an absorbent pad and an adhesive backing pad (Figure 1b). The sample pad was saturated with 0.05 M PB (pH 7.4, containing 0.5% BSA, 0.5% Tween-20, 0.3% PVP and 0.03% ProClin 300) and dried for 12 h at 37 °C. The coating antigen and goat anti-rabbit IgG, which served as the test line and the control line (T line and C line), were diluted with 0.05 M CB (carbonate buffer solution, pH 9.6) and 0.02 M PB (pH 7.4), respectively, to an appropriate concentration, and then sprayed on the NC film with a volume of 0.8 µL/cm. The T line was 8 mm from the bottom of the NC film, and the distance between the T line and the C line was 6 mm. Then the prepared NC membrane was dried at 37 °C for 12 h. Finally, all parts were pasted on a PVC baking card, cut into 3.5 mm-wide strips and placed in a sealed bag with desiccant.

2.7. Sample Preparation

Four slimming health foods with different substrates (capsule, coffee, tea and tablet) were obtained from the local market, and were previously confirmed to be free of furosemide using LC-MS/MS. The outer shell of the capsule was removed to obtain the powder. The coffee, tea and tablets were taken out of their packing bags and ground into powder. An amount of 1.00 g of sample was added into a 10 mL centrifuge tube containing 4 mL of methanol and mixed on a vortex mixer for 2 min. Then the mixture was centrifuged at $4000\times g$ for 3 min. To obtain the sample solution, 200 µL of the supernatant was added to 800 µL of 0.2 M PB (pH 7.4).

2.8. Test Procedure

In this study, the vertical operation mode was used in the strip testing process. We added 150 µL of sample solution and 5 µL of AuNPs-labeled conjugated probe to the microwell, and the probe was gently pipetted back and forth to evenly disperse it in the sample solution. After incubating for 3 min at room temperature, the test strip was inserted immediately and vertically into the microwell. After reacting for another 4 min, the test strip was removed from the microwell. The qualitative result was simply read with the naked eye. The signal intensity of the T line and C line was read and obtained by ImageJ software. In more detail, the optical density of the test zones (negatives and positives) in grayscale mode was measured by the ImageJ software to obtain the color intensity.

2.9. Sensitivity

The cut-off value was utilized to determine the sensitivity of the developed LFIA test strips by assessing the concentration of the furosemide in a series of spiked samples with test strips. The cut-off value of the assay is defined as the furosemide level that causes the T line to disappear completely. The sensitivity in actual samples with different substrates was evaluated separately. Sample preparation was carried out according to the previous description. Blank slimming health food samples were spiked with furosemide standard solution (100 µg/mL, diluted in methanol) to the final concentrations of 0 (control), 0.1, 0.2, 0.4, 0.6, 0.8, 1.0 and 1.2 µg/g. Each level was tested three times ($n = 3$).

2.10. Specificity

To evaluate the specificity of the proposed method, furosemide analogues which may be illegally added to slimming health foods, including hydrochlorothiazide, metolazone, bumetanide, acetazolamide, torasemide and ethacrynic acid, were added to the sample solution (1.2 µg/g) for detection.

2.11. Method Confirmation

Four different substrates of slimming health foods were selected to validate the accuracy of the AuNPs-LFIA. The spiked concentrations were chosen according to the sensitivity of samples of different substrates. The spiked scalar was the furosemide concentration corresponding to 0.25, 0.5, 1 and 2 times the cut-off value. A total of 16 spiked samples were tested. Each sample was combined with 4 mL methanol, mixed on a vortex mixer for 2 min, and then centrifuged at 4000× *g* for 3 min. Half of the supernatant was diluted 5 times with PB and tested with test strips, and the other half was diluted 5 times with methanol and passed through a 0.22 μm filter membrane, then tested with LC-MS/MS. Each sample was tested three times (n = 3).

Chromatographic separation of LC-MS/MS was performed on a Waters CORTECS T3 column (2.1 × 100 mm, 2.7 μm). The column temperature was 30 °C. Mobile phase A was an aqueous solution containing 0.1% formic acid, and phase B was an acetonitrile solution containing 0.1% formic acid, which was used for gradient elution. The sample volume was 1 μL.

2.12. Analysis of Blind Samples

We purchased 16 blind samples of slimming health foods from a local Guangzhou market and analyzed them by the established AuNPs-LFIA and LC-MS/MS. The samples were pretreated following the method above. The established AuNPs-LFIA was used to detect furosemide in the blind samples, and the results were confirmed by LC-MS/MS.

3. Results and Discussion

3.1. Characterization of AuNPs and AuNPs–Abs

It can be seen from the obtained TEM images (Figure 2a) that the prepared AuNPs were monodisperse spherical particles. The diameter of the AuNPs was about 35 nm, which was consistent with the required size. In the stability test of AuNPs, the AuNPs solution stored at 4 °C for 6 months did not appear to aggregate or precipitate (Figure 2b), indicating that its dispersion stability could last for at least six months. According to the results of UV–visible absorption spectrometry (Figure 2c), the AuNPs exhibited maximum absorbance at a wavelength of 534 nm. After combination with antibodies, the maximum absorbance shifted to 543 nm, indicating the formation of the AuNPs–Abs conjugate. The combination of AuNPs and antibodies is generally believed to be the result of electrostatic attraction [34]. Under certain conditions, AuNPs have a negative surface charge, and the antibodies have a positive charge on the surface [35]. From the performance of the zeta potential, when the negatively charged AuNPs were combined with the positively charged antibodies, the result showed a decrease in the AuNPs zeta potential. Figure 2d shows that the zeta potential decreased from 36.4 mV to 18.7 mV after conjugation, which further confirms that the antibodies coupled with AuNPs successfully.

3.2. Optimization

3.2.1. pH Value

The pH values during the reaction were critical to the efficacy of AuNPs–Abs conjugates. Theoretically, the pH of the reaction should be slightly higher than the isoelectric point of the protein. Below the isoelectric point, the antibodies may flocculate and cause the aggregation and precipitation of AuNPs–Abs, which would decrease the accuracy and cause false negatives. Above the isoelectric point, the adsorption effect is limited due to the charge repulsion between the antibodies and the AuNPs, and the color of the test strip would turn light. According to signal intensity and sensitivity, the optimal pH of AuNPs solutions was 7.8, corresponding to 12 μL of 0.2 M K_2CO_3 solution added (Figure 3a).

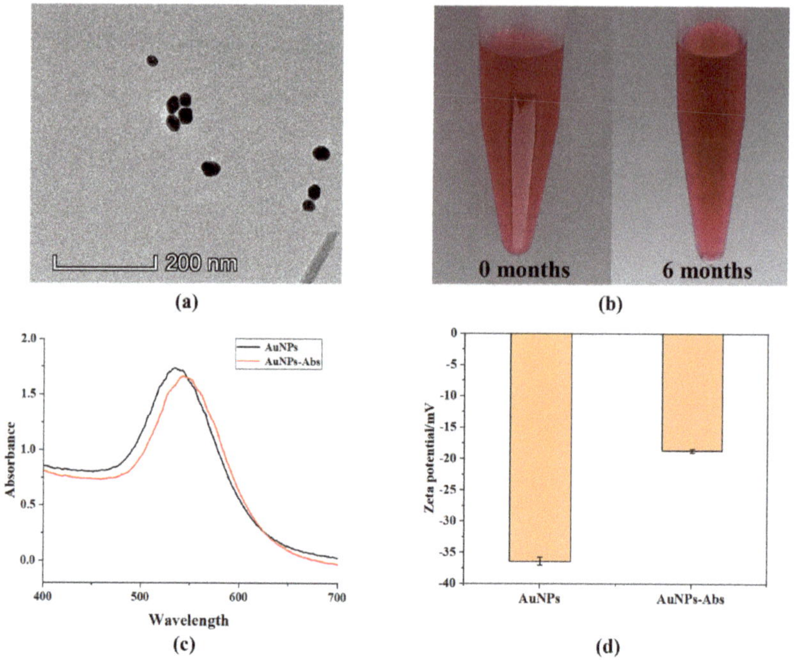

Figure 2. Characterization of Au nanoparticles (AuNPs) and Au nanoparticles-antibodies (AuNPs–Abs). (**a**) Transmission electron microscopy images of AuNPs; (**b**) the results of naked-eye observation of AuNPs at different times; (**c**) the UV–visible absorption spectra of AuNPs and AuNPs–Abs; (**d**) the zeta potential of AuNPs and AuNPs–Abs.

3.2.2. Antibody Concentration

In general, the signal intensity is proportional to the concentration of antibodies, but excess antibodies would affect the sensitivity of the LFIA. In order to screen the optimal concentration of antibodies, antibodies of different concentrations (5, 10, 15 and 20 μg/mL) were added to AuNPs solution to synthesize AuNPs–Abs conjugates. The results show that the highest sensitivity of strip assay was obtained at an antibody concentration of 10 μg/mL (Figure 3b).

3.2.3. Dilution Buffer

The dilution buffer of antigen and goat anti-rabbit second antibody had a great influence on color intensity as a result of the effects on the absorption of protein in the NC membrane caused by the ion type and pH value. In this study, 0.02 M PB (pH 7.4), 0.02 M PBS (phosphate-buffered saline, pH 7.4) and 0.05 M CB (pH 9.6) were selected as the dilution buffers. The results (Figure 3c) show that when the antigen was diluted in 0.05 M CB and the second antibody was diluted in 0.02 M PB, a stronger signal could be obtained.

3.2.4. Resuspension Buffer

The resuspension buffer affected the stability of AuNPs–Abs. In this study, Tris-HCl (pH 7.4), 0.005 M borate buffer solution (pH 8.0), 0.005 M borate buffer solution (pH 8.5) and 0.02 M PB (pH 7.0) were chosen as the resuspension buffers. We found that the resuspension buffers of Tris-HCl and PB could lead to AuNPs–Abs coagulation on the second day, which may have been caused by inappropriate pH values for the dissolution of the AuNPs–Abs probe. In the end, 0.005 M borate buffer solution (pH 8.0), which had the highest assay sensitivity, was chosen as the resuspension buffer (Figure 3d).

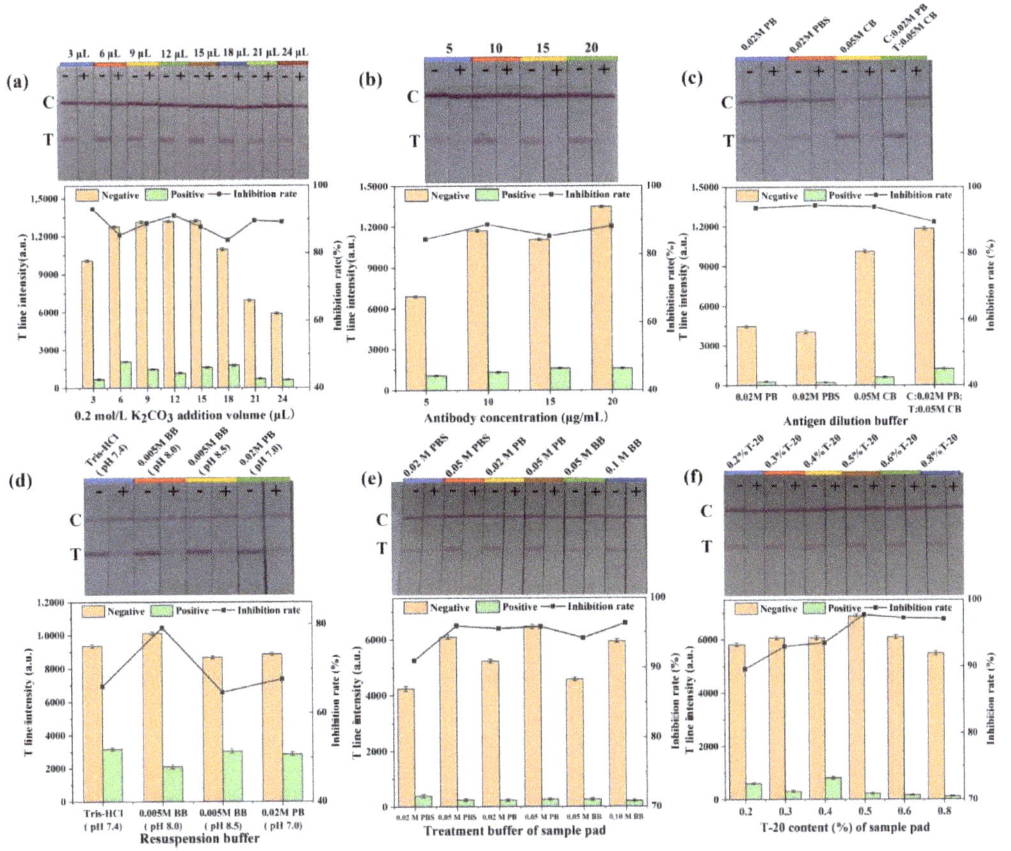

Figure 3. Optimization results of Au nanoparticles lateral flow immunochromatography (AuNPs-LFIA). The LFIA test strip results correspond to the T line intensity obtained by ImageJ software. (**a**) Effect of the pH value represented by 0.2 M K_2CO_3 volume in 1 mL AuNPs; (**b**) antibody concentration of AuNPs–Abs conjugated probe; (**c**) effect of the dilution buffer of coating antigen and second antibody; (**d**) effect of the resuspension buffer of AuNPs–Abs probe; (**e**) effect of ion type and ion concentration in the sample pad treatment solution; (**f**) effect of Tween-20 content in the sample pad treatment solution. −, Negative; +, positive.

3.2.5. Sample Pad Treatment Solution

The sample pad plays a crucial role in reducing the interference of the sample matrix and affected the binding of the labeled probe on the NC membrane, thereby affecting the color intensity and sensitivity of the test strip. Here we mainly evaluated the different buffer and ion concentrations and the Tween-20 content of the sample pad pretreatment solution. As shown in Figure 3e, the higher ion concentration of the sample pad treatment solutions led to better color intensity. In general, Tween-20 improved the fluidity of the sample pad. However, when the flow rate was too fast, it was not conducive to the T-line capture of the AuNPs–Abs. When the flow rate was too slow, it increased the detection time. It can be seen from Figure 3f that as the content of Tween-20 increased, the color intensity of the LFIA first showed an increasing trend, and then decreased. By comparison, we chose a sample pad treatment solution formulation with 0.05 M PB and 0.5% Tween-20 content.

3.3. *Sensitivity*

The qualitative performance of the AuNPs-LFIA was evaluated by the cut-off value. A series of furosemide with different concentrations were spiked into the negative slimming health food samples. Figure 4 shows that the color intensity of T line became weaker with increasing furosemide concentration. When the concentration of furosemide was 0 ng/g, a vigorous color intensity could be seen with the naked eye on the T line, and the color intensity became weaker when a higher concentration of furosemide was added. According to the testing result, the cut-off value was 1.2 μg/g in capsule, coffee and tea samples, and it was 1.0 μg/g in tablet samples. However, the effective dosage of furosemide for an adult is 20–40 mg/day [36]. Therefore, slimming health foods would need to add at least 20 mg of furosemide to the daily dosage to achieve significant weight loss, which is far greater than the LOD of the AuNPs-LFIA. Additionally, compared to the LOD of the HPLC-MS/MS method (2.7 μg/g) established by the Chinese government (BJS 201710) for illegally added furosemide detection in health foods, the established AuNPs-LFIA showed higher sensitivity and achieved on-site detection using a simple operation procedure.

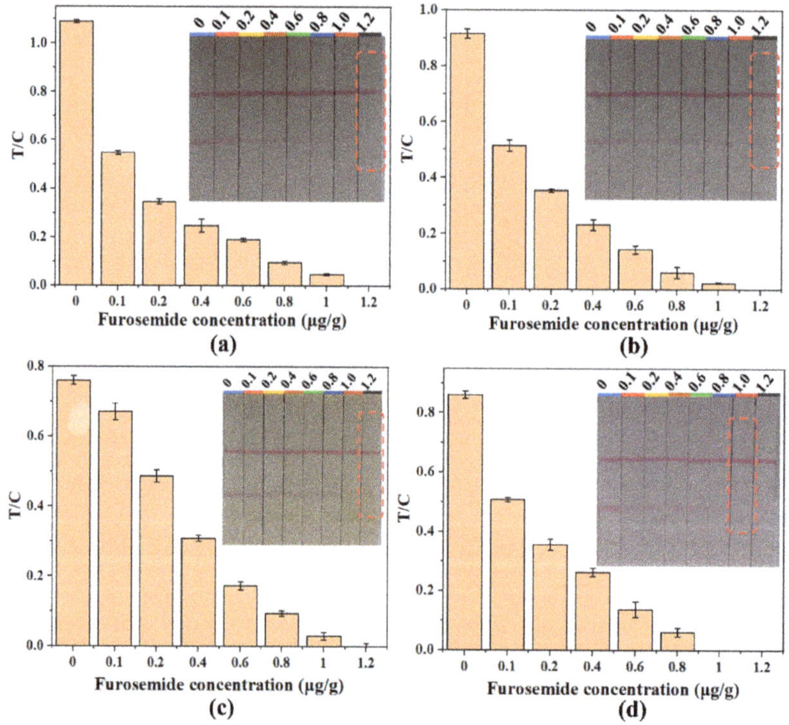

Figure 4. Sensitivity evaluation of the Au nanoparticles lateral flow immunochromatography (AuNPs-LFIA) for furosemide in slimming health foods of different substrates. Red rectangular boxes indicate the color intensity at the cut-off concentrations. Results for determination of furosemide in the slimming capsule samples (**a**), slimming coffee samples (**b**), slimming tea samples (**c**) and slimming tablet samples (**d**).

3.4. Specificity

The developed AuNPs-LFIA was used to detect six furosemide analogues, including hydrochlorothiazide, metolazone, bumetanide, acetazolamide, torasemide and ethacrynic acid at a 1.2 μg/g level. The results showed that the test strip did not detect the other drugs at all, indicating that the LFIA had a high specificity for furosemide detection in slimming health foods (Figure 5).

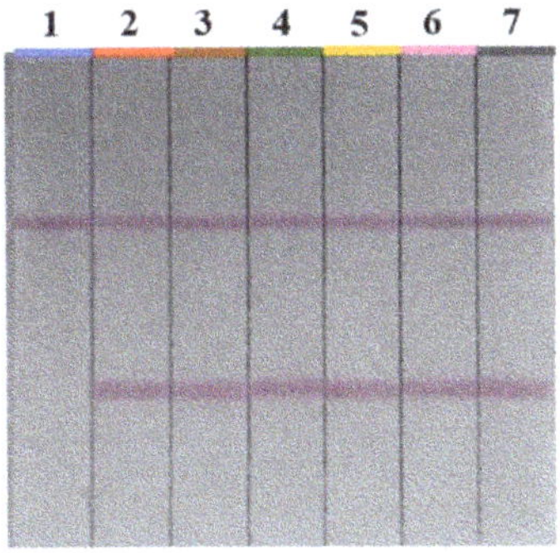

Figure 5. Specificity evaluation results of the Au nanoparticles lateral flow immunochromatography (AuNPs-LFIA): 1, furosemide; 2, hydrochlorothiazide; 3, metolazone; 4, bumetanide; 5, acetazolamide; 6, torasemide; 7, ethacrynic acid.

3.5. Confirmation by LC-MS/MS

LC-MS/MS was employed for the method confirmation. The detection results of AuNPs-LFIA were consistent with LC-MS/MS in all 16 spiked samples. No false-positive or false-negative results were found. The recoveries for furosemide in spiked samples were from 75.83% to 104.53%, with the CVs ranging from 0.09% to 4.92% (Table 1), indicating that the established LFIA is reliable and could be used for large-scale sample screening on-site.

Table 1. Comparison of the detection results of the Au nanoparticles lateral flow immunochromatography (AuNPs-LFIA) and the liquid chromatography with tandem mass spectrometry (LC-MS/MS) in slimming health foods spiked with furosemide (n = 3).

Sample	Spike Level (μg/g)	LC-MS/MS (μg/mL)	Recovery (%)	CV (%)	AuNPs-LFIA Result
coffee	0.30	0.346 ± 0.006	115.39	1.75	−−−
	0.60	0.707 ± 0.009	117.81	1.23	±±±
	1.20	1.379 ± 0.026	114.95	1.89	+++
	2.40	2.781 ± 0.014	115.90	0.49	+++
capsule	0.30	0.302 ± 0.005	100.80	1.56	−−−
	0.60	0.649 ± 0.014	108.10	2.22	±±±
	1.20	1.196 ± 0.059	99.64	4.92	+++
	2.40	2.295 ± 0.013	95.63	0.55	+++
tea	0.30	0.247 ± 0.004	82.25	1.70	−−−
	0.60	0.512 ± 0.016	85.35	3.05	±±±
	1.20	1.048 ± 0.032	87.37	3.02	+++
	2.40	2.127 ± 0.026	88.61	1.22	+++
tablet	0.25	0.200 ± 0.004	80.18	2.09	−−−
	0.50	0.379 ± 0.011	75.83	2.86	±±±
	1.00	0.879 ± 0.006	87.90	0.65	+++
	2.00	1.862 ± 0.002	93.11	0.09	+++

−, Negative; +, positive; ±, weakly positive.

3.6. Analysis of Blind Samples

The 16 slimming health foods purchased in a Guangzhou market were detected by AuNPs-LFIA, and the results showed that all were furosemide-negative samples (Table 2). The results were confirmed and consistent with LC-MS/MS. Although no positive sample was found in this survey due to the sample size and strict supervision in Guangzhou, the results still indicate that the established AuNPs-LFIA is accurate and suitable for the detection of different substrate samples.

Table 2. Comparison of the detection results of the Au nanoparticles lateral flow immunochromatography (AuNPs-LFIA) and the liquid chromatography with tandem mass spectrometry (LC-MS/MS) in slimming health foods purchased in a Guangzhou market ($n = 3$).

Blind Sample	LC-MS/MS	AuNPs-LFIA	Blind Sample	LC-MS/MS	AuNPs-LFIA
 Sample 1	ND	—	 Sample 9	ND	– – –
 Sample 2	ND	—	 Sample 10	ND	– – –
 Sample 3	ND	—	 Sample 11	ND	– – –
 Sample 4	ND	—	 Sample 12	ND	– – –
 Sample 5	ND	—	 Sample 13	ND	– – –
 Sample 6	ND	—	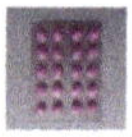 Sample 14	ND	– – –
 Sample 7	ND	—	 Sample 15	ND	– – –
 Sample 8	ND	—	 Sample 16	ND	– – –

ND, Not detected; –, negative.

4. Conclusions

This study developed a sensitive AuNPs-LFIA for the rapid detection of the illegal adulterant drug furosemide that is sometimes found in slimming health foods for the first time. By optimizing a series of parameters that may affect the performance of the AuNPs-LFIA, the sensitivity for furosemide detection was higher than the detection limit of the HPLC-MS/MS method formulated by the Chinese government for the detection of illegal furosemide addition to health food. The sample preparation and test operation of the developed AuNPs-LFIA is 12 min in total, and the procedure is simpler and faster than other existing methods for furosemide detection. In conclusion, the developed AuNPs-LFIA could be applied as an on-site rapid detection method for the screening of furosemide in slimming health foods.

Author Contributions: Conceptualization, H.L. and X.S.; methodology, Y.L.; software, H.X.; validation, H.X. and J.W.; resources, X.L., Z.X. (Zhili Xiao) and Z.X. (Zhenlin Xu); writing—original draft preparation, Y.L.; writing—review and editing, X.S.; project administration, H.L. and X.S.; funding acquisition, H.L. and X.S. All authors have read and agreed to the published version of the manuscript.

Funding: This research was funded by the National Key Research and Development Program of Thirteenth Five-Year Plan (2017YFC1601700), the National Scientific Foundation of China (31871883, 32072316), Guangdong Basic and Applied Basic Research Foundation (2019A1515012107).

Conflicts of Interest: The authors declare no conflict of interest.

References

1. W.H.O. Obesity and Overweight. Available online: https://www.who.int/news-room/fact-sheets/detail/obesity-and-overweight (accessed on 11 August 2021).
2. Yun, J.; Choi, J.; Jo, C.; Kwon, K. Detection of synthetic anti-obesity drugs, designer analogues and weight-loss ingredients as adulterants in slimming Foods from 2015 to 2017. *J. Chromatogr. Sep. Tech.* **2018**, *9*, 2. [CrossRef]
3. Cohen, P.A. The FDA and adulterated supplements—dereliction of duty. *JAMA Netw. Open* **2018**, *1*, e183329. [CrossRef]
4. Muschietti, L.; Redko, F.; Ulloa, J. Adulterants in selected dietary supplements and their detection methods. *Drug Test. Anal.* **2020**, *12*, 861–886. [CrossRef]
5. Eid, P.S.; Ibrahim, D.A.; Zayan, A.H.; Abd Elrahman, M.M.; Shehata, M.A.A.; Kandil, H.; Abouibrahim, M.A.; Luc Minh, D.; Shinkar, A.; Elfaituri, M.K.; et al. Comparative effects of furosemide and other diuretics in the treatment of heart failure: A systematic review and combined meta-analysis of randomized controlled trials. *Heart Fail. Rev.* **2020**, *26*, 127–136. [CrossRef]
6. Silva, E.F.; Tanaka, A.A.; Fernandes, R.N.; Munoz, R.A.A.; Silva, I.S. Batch injection analysis with electrochemical detection for the simultaneous determination of the diuretics furosemide and hydrochlorothiazide in synthetic urine and pharmaceutical samples. *Microchem. J.* **2020**, *157*, 105027. [CrossRef]
7. Banik, M.; Gopi, S.P.; Ganguly, S.; Desiraju, G.R. Cocrystal and Salt Forms of Furosemide: Solubility and Diffusion Variations. *Cryst. Growth Des.* **2016**, *16*, 5418–5428. [CrossRef]
8. Chu, R.; Chen, L.; Zhang, B.; Jia, C.; Qian, Y. Data Analysis of Illegally Adding Chemical Substances in 84 Batches of Slimming Health Foods. *Chin. J. Mod. Appl. Pharm.* **2020**, *16*, 1973–1976. [CrossRef]
9. Cianchino, V.; Acosta, G.; Ortega, C.; Martinez, L.D.; Gomez, M.R. Analysis of potential adulteration in herbal medicines and dietary supplements for the weight control by capillary electrophoresis. *Food Chem.* **2008**, *108*, 1075–1081. [CrossRef] [PubMed]
10. Muller, L.S.; Muratt, D.T.; Dal Molin, T.R.; Urquhart, C.G.; Viana, C.; de Carvalho, L.M. Analysis of Pharmacologic Adulteration in Dietary Supplements by Capillary Zone Electrophoresis Using Simultaneous Contactless Conductivity and UV Detection. *Chromatographia* **2018**, *81*, 689–698. [CrossRef]
11. Moreira, A.P.L.; Motta, M.J.; Dal Molin, T.R.; Viana, C.; de Carvalho, L.M. Determination of diuretics and laxatives as adulterants in herbal formulations for weight loss. *Food Addit. Contam. A* **2013**, *30*, 1230–1237. [CrossRef] [PubMed]
12. Dunn, J.D.; Gryniewicz-Ruzicka, C.M.; Mans, D.J.; Mecker-Pogue, L.C.; Kauffman, J.F.; Westenberger, B.J.; Buhse, L.F. Qualitative screening for adulterants in weight-loss supplements by ion mobility spectrometry. *J. Pharm. Biomed. Anal.* **2012**, *71*, 18–26. [CrossRef]
13. De Carvalho, L.M.; Viana, C.; Moreira, A.P.L.; do Nascimento, P.C.; Bohrer, D.; Motta, M.J.; da Silveira, G.D. Pulsed amperometric detection (PAD) of diuretic drugs in herbal formulations using a gold electrode following ion-pair chromatographic separation. *J. Solid State Electrochem.* **2013**, *17*, 1601–1608. [CrossRef]
14. Muller, L.S.; Dal Molin, T.R.; Muratt, D.T.; Leal, G.C.; Urquhart, C.G.; Viana, C.; de Carvalho, L.M. Determination of Stimulants and Diuretics in Dietary Supplements for Weight Loss and Physical Fitness by Ion-pair Chromatography and Pulsed Amperometric Detection (PAD). *Curr. Anal. Chem.* **2018**, *14*, 562–570. [CrossRef]

15. Zeng, Y.; Xu, Y.M.; Kee, C.L.; Low, M.Y.; Ge, X.W. Analysis of 40 weight loss compounds adulterated in health supplements by liquid chromatography quadrupole linear ion trap mass spectrometry. *Drug Test. Anal.* **2016**, *8*, 351–356. [CrossRef] [PubMed]
16. Muratt, D.T.; Muller, L.S.; Dal Molin, T.; Viana, C.; de Carvalho, L.M. Pulsed amperometric detection of pharmacologic adulterants in dietary supplements using a gold electrode coupled to HPLC separation. *Anal. Methods* **2018**, *10*, 2226–2233. [CrossRef]
17. Rebiere, H.; Guinot, P.; Civade, C.; Bonnet, P.A.; Nicolas, A. Detection of hazardous weight-loss substances in adulterated slimming formulations using ultra-high-pressure liquid chromatography with diode-array detection. *Food Addit. Contam. A* **2012**, *29*, 161–171. [CrossRef] [PubMed]
18. Lancanova Moreira, A.P.; Gobo, L.A.; Viana, C.; de Carvalho, L.M. Simultaneous analysis of antihypertensive drugs and diuretics as adulterants in herbal-based products by ultra-high performance liquid chromatography-electrospray tandem mass spectrometry. *Anal. Methods* **2016**, *8*, 1881–1888. [CrossRef]
19. Said, M.I.; Rageh, A.H.; Abdel-aal, F.A.M. Fabrication of novel electrochemical sensors based on modification with different polymorphs of MnO2 nanoparticles. Application to furosemide analysis in pharmaceutical and urine samples. *RSC Adv.* **2018**, *8*, 18698–18713. [CrossRef]
20. Martins, T.S.; Bott-Neto, J.L.; Raymundo-Pereira, P.A.; Ticianelli, E.A.; Machado, S.A.S. An electrochemical furosemide sensor based on pencil graphite surface modified with polymer film Ni-salen and $Ni(OH)_2/C$ nanoparticles. *Sens. Actuator B Chem.* **2018**, *276*, 378–387. [CrossRef]
21. Heidarimoghadam, R.; Farmany, A. Rapid determination of furosemide in drug and blood plasma of wrestlers by a carboxyl-MWCNT sensor. *Mater. Sci. Eng. C Mater. Biol. Appl.* **2016**, *58*, 1242–1245. [CrossRef]
22. Nagata, S.I.; Kurosawa, M.; Kuwajima, M. A direct enzyme immunoassay for the measurement of furosemide in horse plasma. *J. Vet. Med. Sci.* **2007**, *69*, 305–307. [CrossRef]
23. Woods, W.E.; Wang, C.J.; Houtz, P.K.; Tai, H.H.; Wood, T.; Weckman, T.J.; Yang, J.M.; Chang, S.L.; Blake, J.W.; Tobin, T. Immunoassay detection of drugs in racing horses. VI. Detection of furosemide (Lasix) in equine blood by a one step ELISA and PCFIA. *Res. Commun. Chem. Pathol. Pharmacol.* **1988**, *61*, 111–128. [PubMed]
24. Stanker, L.H.; Muldoon, M.T.; Buckley, S.A.; Braswell, C.; Kamps-Holtzapple, C.; Beier, R.C. Development of a monoclonal antibody-based immunoassay to detect furosemide in cow's milk. *J. Agric. Food Chem.* **1996**, *44*, 2455–2459. [CrossRef]
25. Tang, R.H.; Yang, H.; Choi, J.R.; Gong, Y.; Feng, S.S.; Pingguan-Murphy, B.; Huang, Q.S.; Shi, J.L.; Mei, Q.B.; Xu, F. Advances in paper-based sample pretreatment for point-of-care testing. *Crit. Rev. Biotechnol.* **2017**, *37*, 411–428. [CrossRef]
26. Di Nardo, F.; Chiarello, M.; Cavalera, S.; Baggiani, C.; Anfossi, L. Ten Years of Lateral Flow Immunoassay Technique Applications: Trends, Challenges and Future Perspectives. *Sensors* **2021**, *21*, 5185. [CrossRef]
27. Hnasko, R.M.; Jackson, E.S.; Lin, A.V.; Haff, R.P.; McGarvey, J.A. A rapid and sensitive lateral flow immunoassay (LFIA) for the detection of gluten in foods. *Food Chem.* **2021**, *355*, 129514. [CrossRef] [PubMed]
28. Liu, Y.; Zhan, L.; Qin, Z.; Sackrison, J.; Bischof, J.C. Ultrasensitive and highly specific lateral flow assays for point-of-care diagnosis. *ACS Nano* **2021**, *15*, 3593–3611. [CrossRef] [PubMed]
29. Doyle, J.; Uthicke, S. Sensitive environmental DNA detection via lateral flow assay (dipstick)—A case study on corallivorous crown-of-thorns sea star (Acanthaster cf. solaris) detection. *Environ. DNA* **2021**, *3*, 323–342. [CrossRef]
30. Chen, Z.; Wu, H.; Xiao, Z.; Fu, H.; Shen, Y.; Luo, L.; Wang, H.; Lei, H.; Hongsibsong, S.; Xu, Z. Rational hapten design to produce high-quality antibodies against carbamate pesticides and development of immunochromatographic assays for simultaneous pesticide screening. *J. Hazard. Mater.* **2021**, *412*, 125241. [CrossRef]
31. Lan, J.; Sun, W.; Chen, L.; Zhou, H.; Fan, Y.; Diao, X.; Wang, B.; Zhao, H. Simultaneous and rapid detection of carbofuran and 3-hydroxy-carbofuran in water samples and pesticide preparations using lateral-flow immunochromatographic assay. *Food Agric. Immunol.* **2020**, *31*, 165–175. [CrossRef]
32. Li, J.; Duan, H.; Xu, P.; Huang, X.; Xiong, Y. Effect of different-sized spherical gold nanoparticles grown layer by layer on the sensitivity of an immunochromatographic assay. *RSC Adv.* **2016**, *6*, 26178–26185. [CrossRef]
33. Liu, Z.; Hua, Q.; Wang, J.; Liang, Z.; Li, J.; Wu, J.; Shen, X.; Lei, H.; Li, X. A smartphone-based dual detection mode device integrated with two lateral flow immunoassays for multiplex mycotoxins in cereals. *Biosens. Bioelectron.* **2020**, *158*, 112178. [CrossRef] [PubMed]
34. Christopher, P.; Robinson, N.; Shaw, M.K. Antibody-Label Conjugates in Lateral-Flow Assays. In *Drugs of Abuse: Body Fluid Testing*; Wong, R.C., Tse, H.Y., Eds.; Humana Press: Totowa, NJ, USA, 2005; pp. 87–98. [CrossRef]
35. Kong, D.; Liu, L.; Song, S.; Suryoprabowo, S.; Li, A.; Kuang, H.; Wang, L.; Xu, C. A gold nanoparticle-based semi-quantitative and quantitative ultrasensitive paper sensor for the detection of twenty mycotoxins. *Nanoscale* **2016**, *8*, 5245–5253. [CrossRef] [PubMed]
36. Granero, G.; Longhi, M.; Mora, M.; Junginger, H.; Midha, K.; Shah, V.; Stavchansky, S.; Dressman, J.; Barends, D. Biowaiver monographs for immediate release solid oral dosage forms: Furosemide. *J. Pharm. Sci.* **2010**, *99*, 2544–2556. [CrossRef] [PubMed]

foods

MDPI

Article

Quantitative Determination of Nitrofurazone Metabolites in Animal-Derived Foods Based on a Background Fluorescence Quenching Immunochromatographic Assay

Yuping Wu [1], Jia Wang [2], Yong Zhou [3], Yonghua Qi [1], Licai Ma [4], Xuannian Wang [1,*] and Xiaoqi Tao [2,*]

[1] College of Life Science and Basic Medicine, Xinxiang University, Xinxiang 453003, China; wuyuping62@xxu.edu.cn (Y.W.); qyh@xxu.edu.cn (Y.Q.)
[2] College of Food Science, Southwest University, Chongqing 400715, China; wangjia2020@cau.edu.cn
[3] College of Veterinary Medicine, China Agricultural University, Beijing 100193, China; zhouyong@cau.edu.cn
[4] Beijing WDWK Biotech Co., Ltd., Beijing 100095, China; malicai@wdwkbio.com
* Correspondence: yuefeng@xxu.edu.cn (X.W.); taoxiaoqi@swu.edu.cn (X.T.); Tel.: +86-150-9009-8008 (X.W.); +86-183-0600-8102 (X.T.)

Abstract: Due to their facile synthesis and friendly functionalization, gold nanoparticles (AuNPs) have been applied in all kinds of biosensors. More importantly, these biosensors, with the combination of AuNPs and immunoassay, are expected to be used for the detection of different compounds with low concentrations in complex samples. In this study, a AuNPs-labeled antibody immunoprobe was prepared and combined with a fluorescence-quenching principle and a background fluorescence-quenching immunochromatographic assay (bFQICA), achieving rapid on-site detection. By using a portable fluorescence immunoquantitative analyzer and a QR code with a built-in standard curve, the rapid quantitative determination for nitrofurazone metabolite of semicarbazide (SEM) in animal-derived foods was realized. The limits of detection (LODs) for bFQICA in egg, chicken, fish, and shrimp were 0.09, 0.10, 0.12, and 0.15 μg kg^{-1} for SEM, respectively, with the linear range of 0.08–0.41 μg L^{-1}, the recoveries ranging from 73.5% to 109.2%, and the coefficient of variation <15%, only taking 13 min for the SEM detection. The analysis of animal-derived foods by bFQICA complied with that of liquid chromatography-tandem mass spectrometry (LC-MS/MS).

Keywords: semicarbazide (SEM); background fluorescence-quenching immunochromatographic assay (bFQICA); quantitative determination; animal-derived foods

Citation: Wu, Y.; Wang, J.; Zhou, Y.; Qi, Y.; Ma, L.; Wang, X.; Tao, X. Quantitative Determination of Nitrofurazone Metabolites in Animal-Derived Foods Based on a Background Fluorescence Quenching Immunochromatographic Assay. *Foods* **2021**, *10*, 1668. https://doi.org/10.3390/foods10071668

Academic Editors: Hongtao Lei, Zhanhui Wang and Sergei A. Eremin

Received: 7 June 2021
Accepted: 16 July 2021
Published: 20 July 2021

1. Introduction

As a broad-spectrum antibiotic, nitrofurazone (NFZ) is a well-known member of the nitrofurans class and is widely applied in husbandry to prevent and control a variety of animal diseases caused by *Salmonella* and *Escherichia coli* infection [1,2]. Meanwhile, NFZ was also used as a medicinal feed additive to prevent the dysentery and bacterial enteritis in swine. NFZ as a kind of commonly used drug, can be metabolized to SEM in an animal's body [3]; therefore, the detection of SEM is usually used to reflect the residual state of NFZ. Studies have found that, after a period of accumulation in the human body, nitrofuran metabolites can lead to various organ diseases and can cause serious harm to human health, such as irreversible damage to the central nervous system, liver, kidney, heart, hypothalamus, reproductive system, and so on; toxic and side effects; allergic reaction or allergy; bacterial drug resistance; and dysbacteriosis, teratogenesis, carcinogenesis, and mutagenesis [4,5]. Since 1995, the European Union has prohibited nitrofuran use in livestock, aquaculture, and poultry [6]. Moreover, China and USA have also strictly prohibited nitrofuran application in food-producing animals [7,8]. The European Union and the USA have set the minimum required performance limit (MRPL) (1.0 μg kg^{-1}) for SEM in animal-derived foods [9]. Hence, it is essential to establish effective methods for the detection of SEM in animal-derived foods.

Indeed, various methods have been established for detecting NFZ and/or SEM (the metabolite of NFA) in animal-derived foods, such as high-performance liquid chromatography-ultraviolet (HPLC-UV) [10], HPLC with fluorescence (HPLC-FLD) [11], and HPLC-tandem mass spectrometry (HPLC-MS/MS) [12–16]. However, the above instrumental methods require professional knowledge of operators and costly instruments, and they are unsuitable for on-site detection, which limits their use. The immunoassay is a rapid useful technique for SEM analysis with high throughput tests, such as enzyme-linked immunosorbent assay (ELISA) [17] and fluorescence-linked immunosorbent assay (FLISA) [18]. However, ELISA and FLISA are heterogeneous reactions and time-consuming, which requires tedious washing steps. The colloidal gold immunochromatographic assay (CGICA) [19] is simple, fast, and low cost; however, it always shows the disadvantages of just a qualified determination with a relatively high detection limit.

Due to their facile synthesis and friendly functionalization, gold nanoparticles (AuNPs) have been applied in all kinds of biosensors [20], whether chemical and biological, drug delivery, or photothermal therapy [21–23]. More importantly, these biosensors with the combination of AuNPs and immunoassay are expected to be used for the detection of different compounds with low concentrations in complex samples [24–26]. Wu et al. developed a background fluorescence-quenching immunochromatographic assay (bFQICA) for the detection chloramphenicol (CAP) and aflatoxin M_1 (AFM_1) in milk with the limit of detection (LOD) for CAP of 0.0008 μg L^{-1} and for AFM_1 of 0.0009 μg L^{-1} [25]. In 2020, we successfully conducted the bFQICA to achieve co-determination of quinoxaline-2-carboxylic acid (QCA) and 3-methyl-quinoxaline-2-carboxylic acid (MQCA) in pork, with a sensitivity of 0.1–1.6 μg L^{-1} and only taking 30 min for the detection, exhibiting convenience and efficiency [26]. The bFQICA has the advantages of having high specificity and high sensitivity, and it is quantitative, portable, and accommodates direct read-out mini devices; but so far, there has been no report on SEM detection by the bFQICA.

In this study, a bFQICA, achieving on-site quantitative determination of SEM residues in animal-derived foods (egg, chicken, fish, and shrimp) was established (Figure 1), in which AuNPs were used to quench the fluorescence of a background fluorescence baseboard, and a portable fluorescence immunoquantitative analyzer was used to measure the background fluorescence.

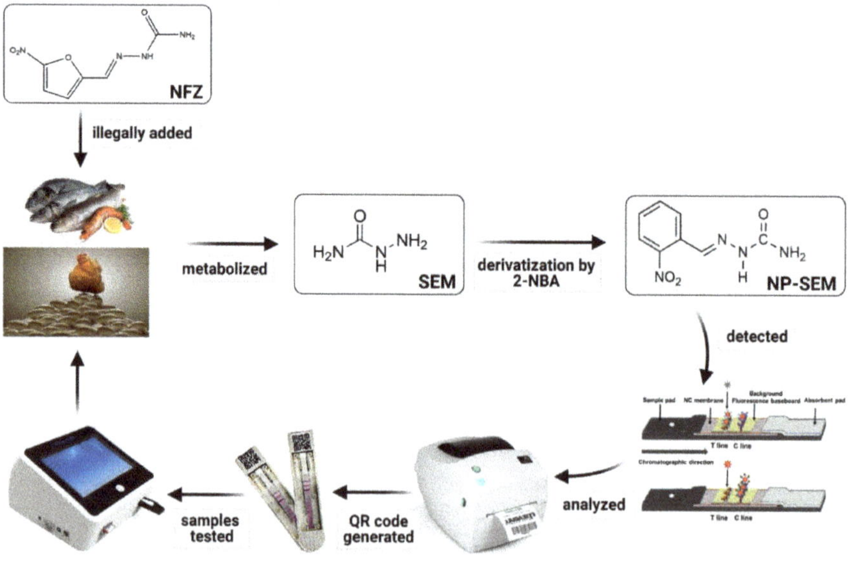

Figure 1. The scheme of bFQICA for detection of SEM in animal-derived foods.

2. Materials and Methods

2.1. Chemicals and Equipment

The parent nitrofurans and SEM were obtained from Dr. Ehrenstorfer (Augsburg, Germany), and other related materials can be seen in the Supplementary Materials. CPSEM-OVA (carboxybenzaldehyde semicarbazone-ovalbumin, 5.47 mg mL^{-1}) and anti-NPSEM monoclonal antibody (mAb) (4.05 mg mL^{-1}) were obtained from Beijing WDWK Biotech Co., Ltd. (Beijing, China). Goat anti-mouse IgG was obtained from Jackson ImmunoResearch Laboratories, Inc. (West Grove, PA, USA).

The sample pad and absorbent pad were from Shanghai Liangxin Co., Ltd. (Shanghai, China). The background fluorescence baseboard was obtained from Shanghai Xinpu Biotechnology Co. Ltd. (Shanghai, China). A fluorescence immune-quantitative analyzer was from Simp Bio-Science Co., Ltd. (Shanghai, China), and the UV-Vis spectrophotometer was obtained from Hitachi Ltd. (Tokyo, Japan). The soft of NiceLabel Pro 2017 was obtained from NiceLabel China (Shanghai, China)

2.2. Preparation and Characterization of AuNPs-Labeled Antibody Immunoprobe

The preparation of the AuNPs-labeled antibody (AuNPs-anti-NPSEM mAb) immunoprobe was according to previous literature with slight modifications [26–28].

First, AuNPs were synthetized by the reduction method of trisodium citrate [29].

Second, for the preparation of AuNPs-anti-NPSEM mAb, the pH of AuNPs (1 mL) was adjusted to 8.0 (K_2CO_3, 0.1 M), then the amount of anti-NPSEM mAb was added, quickly mixed, and incubated for 10 min at room temperature (RT). Afterward, 20 μL of BSA (20%, w/v) was added, mixed for blocking, and the mixture was centrifuged (8000 rpm, 10 min, 4 °C). Finally, the supernatant was quickly moved, and the pellet was diluted in storage buffer (200 μL). In addition, AuNPs-anti-NPSEM mAb (4 μL) was transferred into a microplate well and ultrasonically resuspended, then stored for use (4 °C).

2.3. Preparation of bFQICA Strip

The bFQICA strip contained a sample pad, background fluorescence baseboard, NC membrane, absorbent pad, and background fluorescence baseboard. Initially, CPSEM-OVA was dissolved in 0.02 M PBS with the final concentrations of 0.17 mg/mL and sprayed onto the NC membrane to form test line (T line). Goat anti-mouse IgG was dissolved in PB (0.02 M) with the final concentrations of 0.33 mg/mL and sprayed on the NC membrane as the control line (C line). The spraying amount of CPSEM-OVA and goat anti-mouse IgG was 0.7 μL/cm, with an interval between the T line and C line of 3.00 mm. Then, the as-prepared NC membrane was dried at 45 °C for 2 h. Next, the NC membrane was attached to the fluorescent region of the background fluorescence baseboard; the sample pad and absorbent pad were assembled on the two sides of the background fluorescence baseboard, respectively. Then, on the NC membrane and the assembled background fluorescence baseboard with a 2 mm overlap, the strip was cut into 4.72 mm wide test strips. Finally, all was put into a jam case and assembled into a bFQICA strip, and the assembled strips were stored and kept sealed in a dry environment until use.

2.4. The Procedure of bFQICA for SEM

First, a standard or samples extraction solution (200 μL) was added to the freeze-dried AuNPs-anti-NPSEM mAb immunoprobes, was gently blown by the pipette, and was mixed until the purplish red particles at the bottom of the well were completely dissolved, after which the solution was incubated for 3 min at RT in microplate well. After that, the above mixture (120 μL) was added into the sample pad. As a result, the mixture could move toward the absorbent pad through capillarity. Finally, the strip was measured by the fluorescence immune-quantitative analyzer after 10 min incubation at RT, and the fluorescence signals for (F_1/F_2) T/C lines were measured.

2.5. Standard Curves and Generation of QR-Code

For the quantitative assay, four parameters were input into software (Nice Label Pro 2017) to generate QR-code with the built-in standard curve, and the QR-code was printed by barcode printer (Label Shop). The accurate concentration of analytes could be obtained by scanning the QR-code (Supplementary Materials).

2.6. Sample Pretreatment

The animal-derived foods (egg, chicken, fish, and shrimp) were from Xinxiang local supermarkets and were stored at −20 °C before use. The sample pretreatment was similar to our previous method [26] (Supplementary Materials). Before the detection by bFQICA, the collected solution had a dilution factor of 5, with a sample diluent (0.02 M PBS containing 0.05% Tween-20, pH 7.4) to remove the matrix interference.

2.7. Validation of bFQICA

Because of the low molecular weight of SEM, 2-NBA is often used to derivatize the metabolite to increase the molecular weight in the sample pretreatment process before detection. For validation of bFQICA, animal-derived food samples were confirmed to be SEM-free by LC-MS/MS (Supplementary Materials).

3. Results and Discussion

3.1. Principle of bFQICA for Quantification of SEM

The detection mode of this study was competitive reaction. The background fluorescence of the membrane strip and the relative fluorescence intensity of the T line were detected quantitatively. AuNPs-anti-NPSEM mAb immunoprobes were bound with NPSEM in the standard or samples extraction solution, and then the mixture was dripped onto the sample pad, moving toward the absorbent pad through capillarity. As shown in Figure 2(A1), when there was no (NP)SEM (negative), the immunoprobes (AuNPs-anti-NPSEM mAb) bound with the CPSEM-OVA coated on the T line in the NC membrane, which could obviously quench (cover) the fluorescence of the T line (F_2) generated from the fluorescein of the background fluorescence baseboard. The remaining immunoprobes (AuNPs-anti-NPSEM mAb) continued to move toward the C line and were bound with the goat anti-mouse IgG, generating less fluorescence at the C line (F_1) due to the quenching (covering) of the fluorescein of the background fluorescence baseboard by AuNPs, in which the ratio of F_1/F_2 was maximum (max) (Figure 2(B1)).

Conversely, when (NP)SEM (positive) was present (Figure 2(A2)), the immunoprobes (AuNPs-anti-NPSEM mAb) were bound with the analytes, and then fewer immunoprobes (AuNPs-anti-NPSEM mAb) would bind with the CPSEM-OVA coated on the NC membrane, with less of a quenching (covering) effect, thus generating more fluorescence on the T line (F_2). Moreover, these probes (the unbound immunoprobes (AuNPs-anti-NPSEM mAb) and AuNPs-anti-NPSEM mAb-analytes complex) could be captured by the goat anti-mouse IgG on the C line and an additional quenching (covering) effect occurred, with the less fluorescence of the C line (F_1), in which the ratio of F_1/F_2 was minimum (min) (Figure 2(B2)). As the concentration of (NP)SEM increased, the ratio of F_1/F_2 decreased. Furthermore, F_1 waned and F_2 waxed with the increased concentration of (NP)SEM. In addition, the concentration of (NP)SEM could be directly displayed by the built-in QR-code, which only took 13 min for the detection of (NP)SEM, including 10 min of incubation and 3 min of signal collection and data calculation.

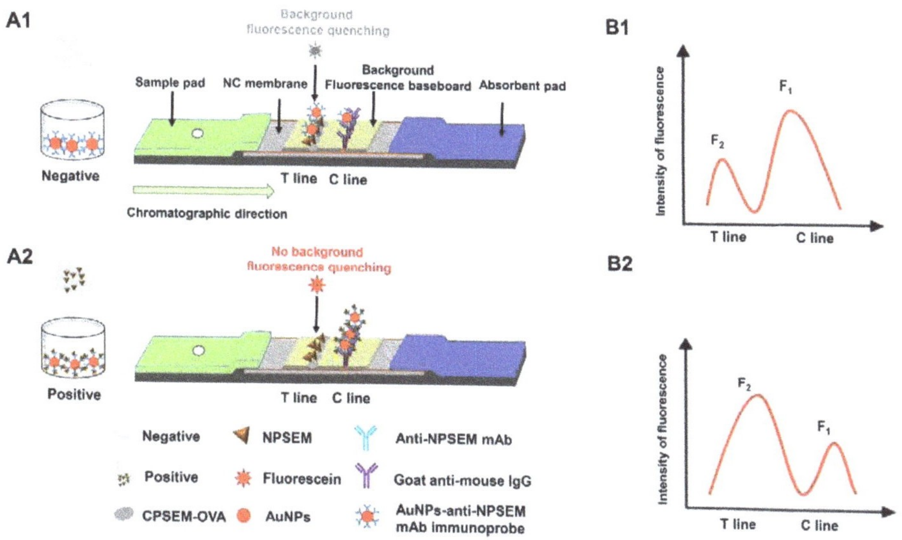

Figure 2. Scheme of bFQICA for the quantitative detection of (NP)SEM. (**A**): The diagram of bFQICA test card, when there was no (NP)SEM (negative) (**A1**), or when there was in the presence of (NP)SEM (positive) (**A2**); (**B**): the fluorescence of C line (F_1) and T line (F_2), when there was no (NP)SEM (negative) (**B1**), or when there was in the presence of (NP)SEM (positive) (**B2**).

3.2. *Characterization of AuNPs*

The solution of the prepared AuNPs was wine red, clear, and uniform, with good dispersibility and no other insoluble impurities, which preliminarily proved that the preparation of AuNPs was successful (Figure 3A).

AuNPs were characterized by UV-Vis spectroscopy with wavelength ranging from 400 to 700 nm, in which the maximum absorption wavelength was 528 nm (Figure 3B), which is the characteristic absorbance peak of AuNPs, indicating a successful preparation. The average diameter of these uniform particles was about 31.5 nm, according to the linear regression equation: $y = 0.4271x + 514.56$ [30], in which y is the maximum wavelength of absorption, and x is the diameter of the gold nanoparticles. The peak width of the maximum absorption peak was narrow and symmetrical, indicating that AuNPs were uniform in size and well dispersed.

The transmission electron microscope of AuNPs is shown in Figure 3C, and the particle size of the AuNPs was about 28–33 nm, consistent with the calculation result of the visible light absorption spectrum of the AuNPs. The results of transmission electron microscope and visible light absorption spectrum showed that the preparation of AuNPs was successful.

3.3. *Optimization and Identification of AuNPs-Labeled Antibody Immunoprobe*

In the preparation process of AuNPs-labeled antibody probe, the particle size of colloidal gold, the amount of antibody, and the pH of the labeling system have great effects on the stability and sensitivity of the AuNPs-labeled antibody probe (Table 1). The scanning results of AuNPs-anti-NPSEM mAb by UV-Vis's spectrophotometer are shown in Figure 3B, whose maximum absorption wavelength had an obvious right shift compared with that of the naked AuNPs. The maximum absorption peak of the AuNPs was 528 nm, and the maximum absorption peaks of the four AuNPs-anti-NPSEM mAb probes was 534.5 nm. The obvious shift of the maximum absorption peak of the AuNPs-anti-NPSEM mAb was due to the increase of the particle size of the antibody adsorbed on the AuNPs surface through electrostatic interaction. At the same time, the maximum absorption peak of the AuNPs-anti-NPSEM mAb probes was narrow and symmetrical, which indicates that the gold labeled antibody probe was stable. This results also verified the successful coupling of the AuNPs-anti-NPSEM mAb probes.

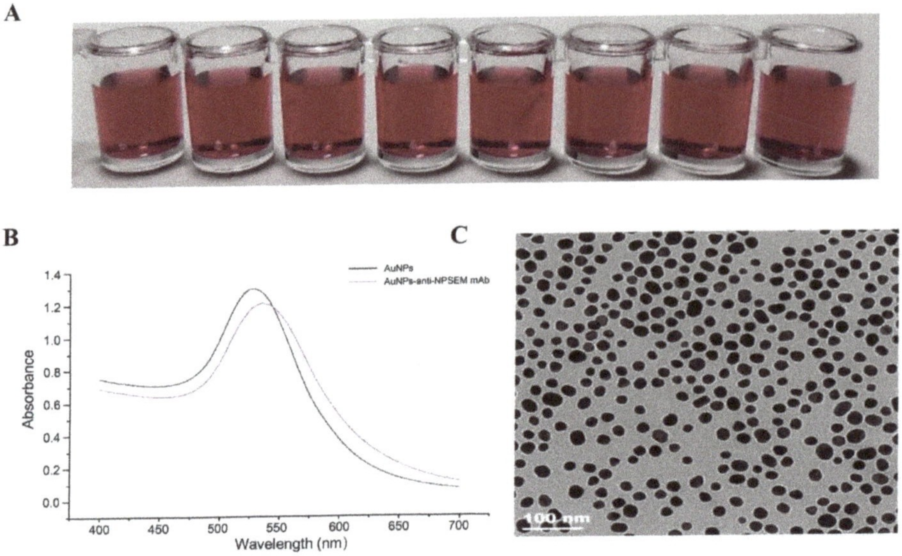

Figure 3. Identification of the prepared AuNPs. (**A**): AuNPs solution in eyes; (**B**): Visible absorption spectrum of naked AuNPs and AuNPs-anti-NPSEM mAb probe, whose concentrations were 4.0 nM and 1.3 mg mL^{-1}, respectively; (**C**): Transmission electron micrograph of AuNPs.

3.4. Optimization of the bFQICA

The concentration of AuNPs-anti-NPSEM mAb probes and the amount of the immunoprobes per strip, and the concentration of coat antigen (CPSEM-OVA) and the goat anti-mouse IgG on the NC membrane were investigated (Table 1). The value of IC_{50} was an important parameter for evaluating the bFQICA performance.

Table 1. Analytical parameters of the bFQICA for the detection of (NP)SEM.

	Characterization	Results
AuNPs-anti-NPSEM mAb probes (1mL reaction system)	The particle size of AuNPs	30 nm
	pH	8 (0.8% v/v K_2CO_3)
	Anti-NPSEM mAb (μg mL^{-1})	2.55
	Storage buffer	0.02 M PB (0.5% BSA, 0.5% Triton X-100, 5% sucrose, 0.03% NaN_3, pH 7.4)
Optimum parameters of the established bFQICA (50 μL reaction system)	The dosage of AuNPs-anti-NPSEM mAb probe	4 μL per well
	AuNPs-anti-NPSEM mAb probe (μg mL^{-1})	2.55
	CPSEM-OVA (mg mL^{-1})	0.17
	Concentration of goat anti-mouse IgG (mg mL^{-1})	0.33
	rehydrated solution (μL)	46 (0.02 PB)
Analytical parameters of NPSEM standard curve	IC_{50} (μg L^{-1})	0.19
	20–80% inhibition (μg L^{-1})	0.08–0.41
	LODs (μg kg^{-1})	0.09 (egg), 0.10 (chicken), 0.12 (fish), 0.15 (shrimp)

3.5. Detectability

SEM was derivatized into NPSEM for detection by the bFQICA. The standard solutions of NPSEM were diluted in PB (0.02 M) to generate the corresponding concentration from 0 to 1.6 μg L^{-1} (0, 0.05, 0.1, 0.2, 0.4, 0.8, 1.6 μg L^{-1}). The standard curves were generated with a series of NPSEM solutions. The detectability of the bFQICA was represented by IC_{50} values of 0.19 μg L^{-1} for NPSEM obtained from the standard curves (Figure 4 and Figure S1). The linear range was 0.08–0.41 μg L^{-1}, represented by the concentrations causing 20–80% inhibition (Table 1).

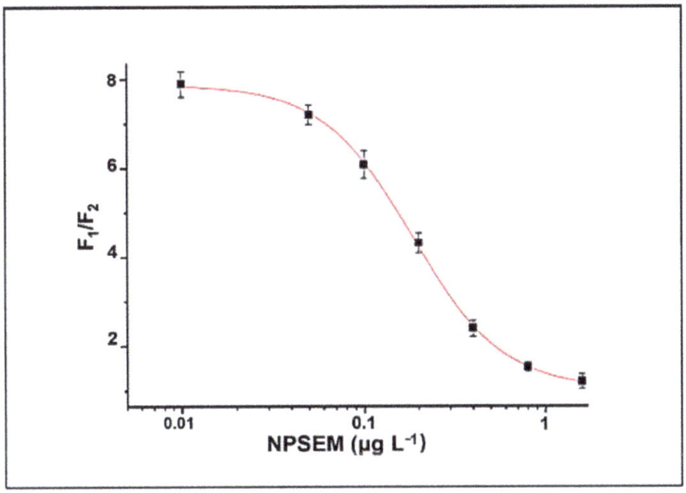

Figure 4. The standard curve of bFQICA method.

In this study, as NFZ was very unstable after entering an animal's body, it could be quickly metabolized into SEM with smaller molecular weight in a short time and could consequently bind to tissue proteins in a relatively stable state. Because the molecular weight of SEM was too small, a derivatization reagent (2-NBA) was usually used to generate NPSEM, increasing its molecular weight [31]. For animal samples, the matrix component with the greatest interference in the extract was protein. Matrix interferences are a common and challenging problem when applying bFQICA to real samples; therefore, sample pretreatment will directly affect the efficiency and accuracy of detection [32]. The purpose of sample pretreatment is to effectively extract, purify, and concentrate the target analyte and reduce the adverse effect of the matrix effect on immune response as much as possible. Generally, the influence of the matrix effect on immunoassay results can be eliminated or weakened by the dilution method [33,34], which can effectively reduce the proportion of non-specific binding. Separation and extraction are also common methods that can eliminate or reduce the matrix effect by removing or reducing matrix components [35]. In this study, on the basis of common sample pretreatment technology (nitrogen blowing method), the amount of derivatization reagent (0.1 mL, 50 mM 2-NBA) was increased appropriately, the temperature of the derivatization process was increased (60 °C) to achieve rapid derivatization [36], and the sample diluent (0.02 M PBS with 0.05% Tween-20, pH 7.4) was prepared to dilute the extract by 5 times, effectively reducing the matrix effect.

3.6. Specificity

There was negligible interference when detecting other chemical substances by the bFQICA (Table S1). The parent nitrofurans, nitrofuran metabolites, and other veterinary drugs commonly used in poultry and aquaculture were individually tested to evaluate the specificity of bFQICA. All the above results indicated the high specificity of the bFQICA for (NP)SEM detection.

3.7. Validation of bFQICA

3.7.1. Limit of Detection

The LODs for bFQICA in egg, chicken, fish, and shrimp were 0.09, 0.10, 0.12, and 0.15 μg kg^{-1} for SEM, respectively. The LOD of the developed bFQICA in egg, chicken, fish, and shrimp were below MRPL of 1.0 μg kg^{-1}, which is compatible with the EU requirements. The bFQICA method not only had the advantages of being a quantitative method for the detection of SEM compared with the published multi-CGICA method [19], but it also had a wider linear range than that of the published MBs-ICA method in fish samples (0.1–50 μg L^{-1}) [37]. Especially, although bFQICA nearly had the detectability of the instrument method using UPLC-MS/MS for SEM [12], it had the advantages of easy operation, low cost, and short implementation time. The developed bFQICA is an improved version of traditional colloidal gold immunochromatography, and it breaks through the bottleneck that AuNPs are usually only suitable for qualitative detection.

3.7.2. Accuracy and Precision

To evaluate the accuracy and precision of the developed bFQICA, blank animal-derived food samples were fortified with SEM at concentrations of LOD, 2LOD, 4LOD, and 1 μg kg^{-1} (MRPL). The recoveries of intra-assay ranged from 75.9% to 104.5%, and the recoveries of inter-assay ranged from 75.7% to 105.1% (Table 2). All the CV values were less than 15%. All the above results confirmed that the bFQICA was an accurate and effective method and that it is fit for the rapid determination of SEM in animal-derived foods.

Table 2. Recovery and precision of SEM added in egg, chicken, fish, and shrimp.

Samples	Spiked Concentration (μg kg^{-1})	Intra-Assay [a]		Inter-Assay [b]	
		Measure ± SD [c] (μg kg^{-1})	Recovery ± CV [d] (%)	Measure ± SD (μg kg^{-1})	Recovery ± CV (%)
Egg	0.09	0.074 ± 0.007	82.2 ± 8.9	0.077 ± 0.008	85.8 ± 9.9
	0.18	0.181 ± 0.012	100.6 ± 6.4	0.168 ± 0.017	93.4 ± 10.3
	0.36	0.302 ± 0.011	83.8 ± 3.8	0.323 ± 0.016	89.6 ± 5.1
	1.00	0.981 ± 0.037	98.1 ± 3.8	0.932 ± 0.049	93.2 ± 5.3
Chicken	0.10	0.090 ± 0.006	90.1 ± 7.2	0.078 ± 0.007	78.1 ± 8.5
	0.20	0.167 ± 0.015	83.5 ± 9.1	0.187 ± 0.012	93.6 ± 6.2
	0.40	0.418 ± 0.015	104.5 ± 3.6	0.420 ± 0.016	105.1 ± 3.9
	1.00	1.012 ± 0.050	101.2 ± 4.9	0.946 ± 0.069	94.6 ± 7.3
Fish	0.12	0.091 ± 0.006	75.9 ± 6.9	0.101 ± 0.006	83.9 ± 6.0
	0.24	0.212 ± 0.012	88.4 ± 5.8	0.211 ± 0.018	88.1 ± 8.6
	0.48	0.461 ± 0.034	96.0 ± 7.4	0.478 ± 0.045	99.6 ± 9.4
	1.00	0.858 ± 0.040	85.8 ± 4.7	0.916 ± 0.057	91.6 ± 6.2
Shrimp	0.15	0.118 ± 0.012	78.4 ± 10.2	0.114 ± 0.013	75.7 ± 11.5
	0.30	0.269 ± 0.017	89.8 ± 6.3	0.279 ± 0.025	92.9 ± 8.9
	0.60	0.555 ± 0.049	92.5 ± 8.8	0.609 ± 0.033	101.5 ± 5.4
	1.00	0.880 ± 0.032	88.0 ± 3.6	0.900 ± 0.060	90.0 ± 6.7

[a] Intra-assay variation was detection by 6 replicates on a single day. [b] Inter-assay variation was detection by 6 replicates on 3 different days. [c] SD, standard deviation. [d] CV, coefficient of variation.

3.8. Application in Field Samples

Eighty field samples of animal-derived food (egg, chicken, fish, and shrimp) were detected by the bFQICA and LC-MS/MS, respectively [14,38]. All the detection results of the two methods were coincident (Table 3), suggesting that the developed bFQICA method was a reliable method for the detection of trace SEM residues in animal-derived foods.

Table 3. Determination of SEM in field animal-derived food samples collected by the bFQICA and LC-MS/MS ($n = 3$).

Sample	No.	bFQICA, Mean ± SD (μg kg^{-1})	LC-MS/MS, Mean ± SD (μg kg^{-1})
Egg	1–9	ND [a]	ND
	10	ND	ND
	11–20	ND	ND
Chicken	1–6	ND	ND
	7	ND	ND
	8–20	ND	ND
Fish	1	0.88 ± 0.04	0.92 ± 0.03
	2–20	ND	ND
Shrimp	1–12	ND	ND
	13	ND	ND
	14–20	ND	ND

[a] ND not detected.

4. Conclusions

This is the first report on the bFQICA method for SEM detection. In this study, the bFQICA for the quantitative determination of SEM in animal-derived foods was successfully developed. The LODs for bFQICA in egg, chicken, fish, and shrimp were 0.09, 0.10, 0.12, and 0.15 μg kg^{-1} for SEM, respectively, with the recoveries ranging from 73.5% to 109.2% (CVs < 15%), using a process that only takes 13 min. The analysis of animal-derived food samples by bFQICA was in accordance with that of LC-MS/MS. Compared with the traditional CGICA method, the detectability of the bFQICA method was higher, and the detection time was shortened compared with heterogeneous reactions such as ELISA. In addition, the concentration of SEM can be directly displayed by the built-in QR-code, which is efficient and convenient. As a promising approach, this method could also be extended for the nitrofurans metabolite in aquaculture and poultry products.

Supplementary Materials: The following are available online at https://www.mdpi.com/article/10.3390/foods10071668/s1, Figure S1: Detection of (NPSEM) of gradient concentration by the bFQICA test card based on grey signal of AuNPs by eyes. Table S1: Cross reactivity (CR) of NPSEM and its analogs by bFQICA test cards.

Author Contributions: Y.W.: conducted the experiments and reviewed and edited the manuscript; J.W., Y.Z., Y.Q., and L.M.: analyzed experimental data and wrote the original draft of the manuscript; X.W. and X.T. contributed to project conceptualization and supervision. All authors have read and agreed to the published version of the manuscript.

Funding: This research was funded by the National Key Research and Development Program of China (2018YFC1602900), Key Scientific Research Projects of Higher Education in Henan Province (19B230015), Doctoral Research Initiation Fund of Xinxiang University (1366020122), Henan Scientific and Technological Project (212102310909).

Institutional Review Board Statement: Not applicable.

Informed Consent Statement: Not applicable.

Conflicts of Interest: The authors declare no conflict of interest. There are no conflicts of interest between the company and the research. The co-author Licai Ma prepared some test strips at Beijing WDWK Biotech Co. using the instruments.

References

1. Vass, M.; Hruska, K.; Franek, M. Nitrofuran antibiotics: A review on the application, prohibition and residual analysis. *Vet. Med.* **2008**, *53*, 469–500. [CrossRef]
2. Liu, H.; Liang, D.P.; Hua, T.G.; She, Z.Y.; Deng, X.F.; Wu, S.Z.; Wu, B. Progress in analytical methods of nitrofuran antibiotics and their metabolites in food: A review. *J. Food Saf. Qual.* **2013**, *4*, 383–388.
3. Kwon, J. Semicarbazide: Natural occurrence and uncertain evidence of its formation from food processing. *Food Control.* **2017**, *72*, 268–275. [CrossRef]

4. Yang, W.Y.; Dong, J.X.; Shen, Y.D.; Yang, J.Y.; Wang, H.; Xu, Z.L.; Yang, X.X.; Sun, Y.M. Indirect competitive chemiluminescence enzyme immunoassay for furaltadone metabolite in Metapenaeus Ensis. *Chin. J. Anal. Chem.* **2012**, *40*, 1816–1821. [CrossRef]
5. Zhao, H.X.; Guo, W.X.; Quan, W.N.; Jiang, J.Q.; Qu, B.C. Occurrence and levels of nitrofuran metabolites in sea cucumber from Dalian, China. *Food Addit. Contam. A* **2016**, *33*, 1672–1677. [CrossRef]
6. European Commission. Regulation (EC) No 470/2009 of the European Parliament and of the council of 6 may 2009. *Off. J. Eur. Union.* **2009**, *L152*, 11–22.
7. Federal, R. Topical nitrofurans, extralabel animal drug use, order of prohibition. *Fed. Reg.* **2002**, *67*, 5470–5471.
8. Announcement No.250, Ministry of Agriculture and Rural Affairs of the People's Republic of China. 2019. Available online: http://extwprlegs1.fao.org/docs/pdf/chn192931.pdf (accessed on 8 July 2021). (In Chinese)
9. European Commission. 2003/181/EC Commission decision of 13 march amending decision 2002/657/EC. *Off. J. Eur. Union.* **2003**, *L71*, 17–18.
10. Tang, T.; Wei, F.D.; Wang, X.; Ma, Y.J.; Song, Y.Y.; Ma, Y.S.; Song, Q.; Xu, G.H.; Cen, Y.; Hu, Q. Determination of semicarbazide in fish by molecularly imprinted stir bar sorptive extraction coupled with high performance liquid chromatography. *J. Chromatog. B* **2018**, *1076*, 8–14. [CrossRef] [PubMed]
11. Sheng, L.Q.; Chen, M.M.; Chen, S.S.; Du, N.N.; Liu, Z.D.; Song, C.F.; Qiao, R. High-performance liquid chromatography with fluorescence detection for the determination of nitrofuran metabolites in pork muscle. *Food Addit. Contam. A* **2013**, *30*, 2114–2122. [CrossRef] [PubMed]
12. Aldeek, F.; Hsieh, K.C.; Ugochukwu, O.N.; Gerard, G.; Hammack, W. Accurate quantitation and analysis of nitrofuran metabolites, chloramphenicol, and florfenicol in seafood by ultrahigh-performance liquid chromatography-tandem mass spectrometry: Method validation and regulatory samples. *J. Agric. Food Chem.* **2018**, *66*, 5018–5030. [CrossRef]
13. Zhang, Y.; Qiao, H.; Chen, C.; Wang, Z.; Xia, X. Determination of nitrofurans metabolites residues in aquatic products by ultra-performance liquid chromatography-tandem mass spectrometry. *Food Chem.* **2016**, *192*, 612–617. [CrossRef]
14. Zhang, Z.W.; Wu, Y.P.; Li, X.W.; Wang, Y.Y.; Li, H.; Fu, Q.; Shan, Y.W.; Liu, T.H.; Xia, X. Multi-class method for the determination of nitroimidazoles, nitrofurans, and chloramphenicol in chicken muscle and egg by dispersive-solid phase extraction and ultra-high performance liquid chromatography-tandem mass spectrometry. *Food Chem.* **2017**, *217*, 182–190. [CrossRef] [PubMed]
15. Valera-Tarifa, N.M.; Plaza-Bolaños, P.; Romero-González, R.; Martínez-Vidal, J.L.; Garrido-Frenich, A. Determination of nitrofuran metabolites in seafood by ultra-high performance liquid chromatography coupled to triple quadrupole tandem mass spectrometry. *J. Food Compos. Anal.* **2013**, *30*, 86–93. [CrossRef]
16. Kulikovskii, A.V.; Gorlov, I.F.; Slozhenkina, M.I.; Vostrikova, N.L.; Kuznetsova, O.A. Determination of nitrofuran metabolites in muscular tissue by high-performance liquid chromatography with mass spectrometric detection. *J. Anal. Chem.* **2019**, *74*, 906–912. [CrossRef]
17. Jiang, W.X.; Luo, P.J.; Wang, X.; Chen, X.; Zhao, Y.F.; Shi, W.; Wu, X.P.; Wu, Y.N.; Shen, J.Z. Development of an enzyme-linked immunosorbent assay for the detection of nitrofurantoin metabolite, 1-amino-hydantoin, in animal tissues. *Food Control.* **2012**, *23*, 20–25. [CrossRef]
18. Sun, Q.; Luo, J.H.; Zhang, L.; Zhang, Z.H.; Le, T. Development of monoclonal antibody-based ultrasensitive enzyme-linked immunosorbent assay and fluorescence-linked immunosorbent assay for 1-aminohydantoin detection in aquatic animals. *J. Pharmaceut. Biomed.* **2018**, *147*, 417–424. [CrossRef] [PubMed]
19. Wang, Q.; Liu, Y.; Wang, M.; Chen, Y.; Jiang, W. A multiplex immunochromatographic test using gold nanoparticles for the rapid and simultaneous detection of four nitrofuran metabolites in fish samples. *Anal. Bioanal. Chem.* **2018**, *410*, 223–233. [CrossRef]
20. Sharma, D.; Lee, J.; Seo, J.; Shin, H.J.S. Development of a sensitive electrochemical enzymatic reaction-based cholesterol biosensor using nano-sized carbon interdigitated electrodes decorated with gold nanoparticles. *Sensors* **2017**, *17*, 2128. [CrossRef]
21. Dykmana, L.; Khlebtsov, N. Gold Nanoparticles in Biomedical Applications: Recent Advances and Perspectives. *Chem. Soc. Rev.* **2012**, *41*, 2256–2282. [CrossRef]
22. Dreaden, E.; Alkilany, A.; Huang, X.; Murphy, C.; El-Sayed, M. The Golden Age: Gold Nanoparticles for Biomedicine. *Chem. Soc. Rev.* **2012**, *41*, 2740–2779. [CrossRef]
23. Grzelczak, M.; Liz-Marzan, L.; Klajn, R. Stimuli-Responsive Self-Assembly of Nanoparticles. *Chem. Soc. Rev.* **2019**, *48*, 1342–1361. [CrossRef] [PubMed]
24. Chen, X.J.; Xu, Y.Y.; Yu, J.S.; Li, J.T.; Zhou, X.L.; Wu, C.Y.; Ji, Q.L.; Ren, Y.; Wang, L.Q.; Huang, Z.Y.; et al. Antigen detection based on background fluorescence quenching immunochromatographic assay. *Anal. Chim. Acta.* **2014**, *841*, 44–50. [CrossRef] [PubMed]
25. Wu, X.; Tian, X.; Xu, L.; Li, J.; Li, X.; Wang, Y. Determination of aflatoxin m1 and chloramphenicol in milk based on background fluorescence quenching immunochromatographic assay. *BioMed Res. Int.* **2017**, *2017*, 8649314. [CrossRef] [PubMed]
26. Wan, X.L.; Wan, X.; Tao, X.Q. Determination of 3-methyl-quinoxaline-2-carboxylic acid and quinoxaline-2-carboxylic acid in pork based on a background fluorescence quenching immunochromatographic assay. *Anal. Sci.* **2020**, *36*, 783–785. [CrossRef]
27. Zhou, Y.; Pan, F.; Li, Y.; Zhang, Y.; Zhang, J.; Lu, S.; Ren, H.; Liu, Z. Colloidal Gold Probe-Based Immunochromatographic Assay for the Rapid Detection of Brevetoxins in Fishery Product Samples. *Biosens. Bioelectron.* **2009**, *24*, 2744–2747. [CrossRef]
28. Wu, W.; Li, J.; Pan, D.; Li, J.; Song, S.; Rong, M.; Li, Z.; Gao, J.; Lu, J. Gold Nanoparticle-Based Enzyme-Linked Antibody-Aptamer Sandwich Assay for Detection of Salmonella Typhimurium. *ACS Appl. Mater. Inter.* **2014**, *6*, 16974–16981. [CrossRef] [PubMed]
29. Sun, X.; Zhao, X.; Jian, T.; Zhou, J.; Chu, F.S. Preparation of gold-labeled antibody probe and its use in immunochromatography assay for detection of aflatoxin B1. *Int. J. Food Microbiol.* **2005**, *99*, 185–194.

30. Peng, J.C.; Liu, X.D.; Ding, X.P.; Fu, Z.J.; Wang, Q.L. Evaluation of the particle diameter of colloidal gold and its distribution through visible spectroscopy. *Bull. Acad. Mil. Med. Sci.* **2000**, *24*, 211–212.
31. Wang, K.; Kou, Y.; Wang, M.; Ma, X.; Wang, J. Determination of nitrofuran metabolites in fish by ultraperformance liquid chromatography-photodiode array detection with thermostatic ultrasound-assisted derivatization. *ACS Omega* **2020**, *5*, 18887–18893. [CrossRef]
32. Points, J.; Burns, D.T.; Walker, M.J. Forensic issues in the analysis of trace nitrofuran veterinary residues in food of animal origin. *Food Control.* **2015**, *50*, 92–103. [CrossRef]
33. Berlina, A.N.; Taranova, N.A.; Zherdev, A.V. Quantum dot-based lateral flow immunoassay for detection of chloramphenicol in milk. *Anal. Bioanal. Chem.* **2013**, *405*, 4997–5000. [CrossRef]
34. Wang, L.; Wang, S.; Zhang, J.Y.; Liu, J.W.; Zhang, Y. Enzyme-linked immunosorbent assay and colloidal gold immunoassay for sulphamethazine residues in edible animal foods: Investigation of the effects of the analytical conditions and the sample matrix on assay performance. *Anal. Bioanal. Chem.* **2008**, *390*, 1619–1627. [CrossRef] [PubMed]
35. Mitchell, J.S.; Lowe, T.E. Matrix effects on an antigen immobilized format for competitive enzyme immunoassay of salivary testosterone. *J. Immunol. Methods* **2009**, *349*, 61–66. [CrossRef]
36. Cooper, K.M.; Caddell, A.; Elliott, C.T.; Kennedy, D.G. Production and characterisation of polyclonal antibodies to a derivative of 3-amino-2-oxazolidinone, a metabolite of the nitrofuran furazolidone. *Anal. Chim. Acta* **2004**, *520*, 79–86. [CrossRef]
37. Lu, X.W.; Liang, X.L.; Dong, J.H.; Fang, Z.Y.; Zeng, L.W. Lateral flow biosensor for multiplex detection of nitrofuran metabolites based on functionalized magnetic beads. *Anal. Bioanal. Chem.* **2016**, *408*, 6703–6709. [CrossRef]
38. Tao, Y.F.; Chen, D.M.; Wei, H.M.; Pan, Y.H.; Liu, Z.L.; Huang, L.L.; Wang, Y.L.; Xie, S.Y.; Yuan, Z.H. Development of an accelerated solvent extraction, ultrasonic derivatisation LC-MS/MS method for the determination of the marker residues of nitrofurans in freshwater fish. *Food Addit. Contam. Part A* **2012**, *29*, 736–745. [CrossRef] [PubMed]

foods

MDPI

Article

Antibody Generation and Rapid Immunochromatography Using Time-Resolved Fluorescence Microspheres for Propiconazole: Fungicide Abused as Growth Regulator in Vegetable

Bo Chen [1], Xing Shen [1], Zhaodong Li [2], Jin Wang [1], Xiangmei Li [1], Zhenlin Xu [1], Yudong Shen [1], Yi Lei [3], Xinan Huang [4], Xu Wang [5] and Hongtao Lei [1,*]

1 Guangdong Province Key Laboratory of Food Quality and Safety, College of Food Science, South China Agricultural University, Guangzhou 510642, China; 13971356810@163.com (B.C.); shenxing325@163.com (X.S.); wangjin940810@stu.scau.edu.cn (J.W.); lixiangmei12@163.com (X.L.); xzlin@scau.edu.cn (Z.X.); shenyudong@scau.edu.cn (Y.S.)
2 College of Materials and Energy, South China Agricultural University, Guangzhou 510642, China; scaulizhaodong@scau.edu.cn
3 Guangdong Institute of Food Inspection, Zengcha Road, Guangzhou 510435, China; Leiy04@foxmail.com
4 Tropical Medicine Institute and South China Chinese Medicine Collaborative Innovation Center, Guangzhou University of Chinese Medicine, Guangzhou 510405, China; xahuang@chinmednetworks.org
5 Institute of Quality Standard and Monitoring Technology for Agro-Products of Guangdong Academy of Agricultural Sciences, Guangzhou 510405, China; wangxuguangzhou@126.com
* Correspondence: hongtao@scau.edu.cn; Tel.: +86-20-8528-3925; Fax: +86-20-8528-0270

Citation: Chen, B.; Shen, X.; Li, Z.; Wang, J.; Li, X.; Xu, Z.; Shen, Y.; Lei, Y.; Huang, X.; Wang, X.; et al. Antibody Generation and Rapid Immunochromatography Using Time-Resolved Fluorescence Microspheres for Propiconazole: Fungicide Abused as Growth Regulator in Vegetable. *Foods* **2022**, *11*, 324. https://doi.org/10.3390/foods11030324

Academic Editor: Evaristo Ballesteros

Received: 14 November 2021
Accepted: 21 January 2022
Published: 24 January 2022

Abstract: Propiconazole (PCZ) is a fungicide popularly used to prevent and control wheat and rice bakanae disease, etc. However, it was recently found to be illegally employed as a plant regulator to induce thick stems and dark green leaves of *Brassica campestris*, a famous vegetable in Guangdong, South China. Due to a lack of available recognition molecules to the target analyte, it is still a big challenge to establish a rapid surveillance screening method. In this study, a novel chiral hapten was rationally designed, and an artificial immunogen was then prepared for the generation of a specific antibody against propiconazole for the first time. Using the obtained antibody, a highly sensitive time-resolved fluorescence microspheres lateral flow immunochromatographic assay (TRFMs-LFIA) was established with a visual limit of detection of 100 ng/mL and a quantitative limit of detection of 1.92 ng/mL for propiconazole. TRFMs-LFIA also exhibited good recoveries ranging from 78.6% to 110.7% with coefficients of variation below 16%. The analysis of blind real-life samples showed a good agreement with results obtained using HPLC-MS/MS. Therefore, the proposed method could be used as an ideal screening surveillance tool for the detection of propiconazole in vegetables.

Keywords: propiconazole; hapten; antibody; time-resolved fluorescence; lateral flow immunochromatographic assay

1. Introduction

Propiconazole (PCZ) is a systemic triazole fungicide that can prevent most fungal diseases in banana, wheat and rice [1]. It is favored by farmers for its broad-spectrum sterilization and long duration. The crops registered for using of propiconazole in China are banana, wheat and rice [2]. In 2018, the annual use of propiconazole in China reached 2087.42 tons, which exceeded other countries [3]. In the hazard classification, propiconazole is classified as reproductive toxicity 1B category and belongs to endocrine disrupting substances [4]. Related studies have confirmed that propiconazole can result in genetic toxicity, liver toxicity and growth toxicity [5–7]. The U.S. Environmental Protection Agency has included propiconazole in the list of possible human carcinogens [8].

Most Cantonese prefer brassica campestris with thick stems and a dark green color, which may be the most favored vegetable by local people in Guangdong due to its rich nutrition and taste [9]. However, in order to cater to the preferences of Cantonese, propiconazole was sometimes illegally used as a plant growth regulator to obtain a good appearance of brassica campestris [10]. During the first three quarters of 2016, an average of 0.011~0.07 mg/kg propiconazole was detected brassica campestris samples from 11 farmer markets and supermarkets in Guangzhou (capital city of Guangdong province, China) [11]. Based on 84 brassica campestris samples from nine farmer markets in Guangzhou, the average residue of propiconazole was 0.06 mg/kg, and the average positive detection rate was as high as 90% [12]. All these indicate propiconazole residues are commonly found in brassica campestris in Guangzhou markets.

Japan stipulates that the maximum residue of propiconazole in brassica and leafy vegetables is 0.05 mg/kg [13]. The EU cancelled the registration of propiconazole in 2019 due to a lack of data on the toxicity evaluation of propiconazole metabolites, and the existing data could not complete the risk assessment related to consumer dietary intake [14]. According to the China food safety standard "Maximum Residue Limits of Pesticides in Foods" (GB 2763-2019), some vegetables and fruits have been set a maximum residue limit for propiconazole residues, e.g., typhalatifolia (0.05 mg/kg) and banana (1 mg/kg). However, there is no residue limit set for propiconazole in many other vegetables.

Methods for detecting propiconazole include gas chromatography (GC) [15,16], high-performance liquid chromatography (HPLC) [17,18], gas chromatography-tandem mass spectrometry (GC-MS/MS) [19,20], and liquid chromatography-tandem mass spectrometry (LC-MS/MS) [21,22]. These methods are widely used due to their advantages in sensitivity and accuracy [23]. However, they are complex, require professionals to operate, and are time- and cost-sensitive [24]. The immunoassay has become the most important part of the rapid detection field because of its time-saving, high sensitivity, and simple operation [24]. Until now, there are limited reported immunoassays for propiconazole detection, mainly colloidal gold immunochromatographic assay (CG-ICA) [25] and indirect competitive enzyme-linked immunosorbent assay (ic ELISA) [26]. Nevertheless, the ELISA method requires laboratory conditions and tedious washing steps. In order to establish a highly sensitive as well as convenient propiconazole immunoassay, fluorescent microspheres-based immunochromatography has great potential.

In this study, a portable and rapid immunochromatographic assay using time-resolved fluorescent microspheres as a tracer was established for the detection of propiconazole. The hapten used in the immunochromatographic assay was rationally designed using a similar chiral carbon containing structure to that of propiconazole, and a polyclonal antibody that specifically recognizes propiconazole was obtained. Based on this antibody, a time-resolved fluorescence microspheres lateral flow immunochromatographic assay (TRFMs-LFIA) was developed, optimized, and evaluated for its sensitivity, specificity, and recovery, and then applied for the analysis of blind vegetable samples.

2. Materials and Methods

2.1. Reagents and Solutions

Propiconazole (PCZ), difenoconazole, diniconazole, hexaconazole, tebuconazole, epoxiconazole, myclobutanil, paclobutrazol, flusilazole, cyproconazole, triadimenol, bitertanol Standard, triadimefon, N,N-Dimethylformamide (DMF), N-hydroxysuccinimide (NHS), 1-ethyl-3-(3-dimethylaminopropyl)-carbodiimide (EDC), ovalbumin (OVA), keyhole limpet hemocyanin (KLH), bovine serum albumin (BSA), graphitized carbon black (GCB), N-Propylethylenediamine (PSA), 4-Dimethylaminopyridine (DMAP), N,N-Carbonyldiimidazole (CDI), succinic anhydride anhydrous dichloromethane, and 2-(N-morpholino) ethanesulfonic acid (MES) were purchased from Sigma-Aldrich (St. Louis, MO, USA). ((2S,4S)-2-((1H-1,2,4-triazol-1-yl) methyl)-2-(2,4-dichlorophenyl)-1,3-dioxolan-4-yl) methanol (AZC) was purchased from Toronto Research Chemicals (Toronto, Canada). Time-resolved fluorescence microsphere (TRFM), europium chelate (365/610), 0.2 μm, 1% (*w/v*) solid suspension, was purchased from Bangs Laboratories Inc. (Indiana, USA). Nitrocellulose filter membrane (Sartorius, UniSart CN95) were obtained from Sartorius Stedim Biotech GmbH (Goettingen, Germany). Sample pad (GF-2), absorbent pad (CH37 K), adhesive backing pad (SMA31-40), and goat anti-rabbit-immunoglobulin G (IgG) were purchased from Shanghai Liangxin Co. Ltd. (Shanghai, China). Microtiter plates were obtained from the Guangzhou JET BIOFIL Co. (Guangzhou, China). Ultra-pure water was produced using a Unique R-10 water purification system (Unique R-10, Bedford, MA, USA). Chloroauric acid, trisodium citrate, polyvinyl pyrrolidone (PVP), and other chemical substances were purchased from Sinopharm Chemical Reagent Co., Ltd. (Shanghai, China).

New Zealand White Rabbit were purchased from the Guangdong Medical Experimental Animal Centre and raised at the Animal Experiment Centre of South China Agriculture University (Animal Experiment Ethical Approval Number: 2020009, Figure S1). All required licenses were secured prior to commencement of the animal experiments.

2.2. *Apparatus*

The BioDot-XYZ 3060 Dispensing Platform was supplied by BioDot, Inc, (Irvine, CA, USA). The programmable strip cutter ZQ-2000 was purchased from Shanghai kinbio Tech. Co., Ltd. (Shanghai, China). The Lynx 4000 centrifuge was supplied by Thermo Fisher Scientific GmbH (Berlin, Germany). An FIC-Q1 multifunctional fluorescence reader was purchased from Fenghang technology Co., Ltd. (Hangzhou, China). The Nano Drop 2000C ultra-violet spectrophotometer was supplied by Thermo Scientific (Waltham, MA, USA). The UV spectrometer was purchased from Qiangyun Co. (Shanghai, China). The Zetasizer Nano ZS90 used for measurements of size and charge of nanoparticles was supplied by Malvern Panalytical (Malvern, UK). Agilent 1290-6470 Liquid Mass Spectrometry (USA AB SCIEX).

2.3. *Synthesis of Hapten AZC-HS*

The scheme for the hapten synthesis is shown in Figure 1A. Briefly, AZC (0.5 g, 1.5 mmol) was dissolved in 10 mL of anhydrous dichloromethane in a flask. After adding succinic anhydride (0.3 g, 3 mmol) and DMAP (0.018 g, 0.15 mmol), the mixture was stirred at room temperature for 12 h. Then, the mixture was evaporated to dryness before introducing water, extracted with ethyl acetate, and the organic phase was separated, dried over anhydrous Na_2SO_4, and finally concentrated to afford the hapten, 4-(((2S,4R)-2-((1H-1,2,4-triazol-1-yl) methyl)-2-(2,4-dichlorophenyl)-1,3-dioxolan-4-yl) methoxy)-4-oxobutanoic acid (AZC-HS). ESI-MS analysis (negative): m/z 428.8 $[M-H]^-$; ^{1}H NMR (600 MHz, DMSO-d6) δ 12.28–12.23 (m, 1H), 8.40 (d, J = 5.2 Hz, 2H), 7.84 (d, J = 5.1 Hz, 2H), 7.66 (d, J = 5.3 Hz, 2H), 7.46 (t, J = 6.7 Hz, 2H), 7.41 (q, J = 10.2, 7.0 Hz, 2H), 4.84–4.74 (m, 4H), 4.23 (q, J = 5.6 Hz, 2H), 3.98 (dd, J = 10.8, 5.6 Hz, 2H), 3.85 (ddd, J = 26.8, 12.4, 5.9 Hz, 5H), 3.65 (dt, J = 11.7, 5.7 Hz, 2H), 2.55 (t, J = 6.0 Hz, 4H), 2.50 (q, J = 6.6, 6.2 Hz, 6H), 2.42 (d, J = 5.3 Hz, 1H), 1.23 (d, J = 5.4 Hz, 1H).

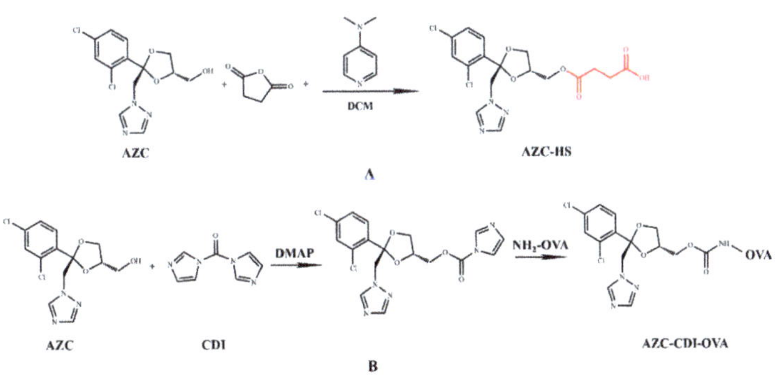

Figure 1. (**A**) Synthesis of the AZC-HS hapten. (**B**) Hapten-carrier conjugation of AZC-CDI-OVA.

2.4. Preparation of Immunogen and Coating Antigen

The immunogen was synthesized using an active ester method [27] with slight modifications. Briefly, 20 mg of hapten AZC-HS was dissolved in 3 mL of DMF. Then, 12.84 mg of EDC and 10 mg of NHS were added and kept stirring at room temperature for 5 h. The mixture was added dropwise to a reaction flask containing KLH (10 mg/mL, 1 mL), and stirred at 4 °C for 12 h. The reaction mixture was purified by dialysis against PBS (0.01 M, pH 7.4) for 3 days to remove the non-reacted reactants. The dialyzed product was the immunogen (AZC-HS-KLH). Full wavelength ultraviolet-visible (UV-Vis) spectroscopy scan was used to confirm the conjugation of the immunogen (Figure S2A), which was finally stored at −20 °C until use.

Coating antigens were prepared using a CDI method [28], OVA was used as the carrier (Figure 1B). Briefly, 110 mg of AZC was dissolved in 3 mL of DMF. Then, 15 mg of CDI and 5 mg of DMAP were added with stirring at room temperature for 24 h. After that, the mixture was added to dropwise to a reaction flask containing OVA (10 mg/mL, 1 mL), and was stirred at 4 °C for 12 h. The reactive mixture was also dialyzed against PBS (0.01 M, pH 7.4) at 4 °C for 3 days. The dialyzed product was the coating antigen (AZC-CDI-OVA) (Figure S2B) and confirmed by UV-Vis scan as above before being stored at −20 °C until use.

AZC, ((2S,4S)-2-((1H-1,2,4-triazol-1-yl) methyl)-2-(2,4-dichlorophenyl)-1,3-dioxolan-4-yl) methanol.

AZC-HS, 4-(((2S,4R)-2-((1H-1,2,4-triazol-1-yl) methyl)-2-(2,4-dichlorophenyl)-1,3-dioxolan-4-yl) methoxy)-4-oxobutanoic acid.

OVA, Albumin from chicken egg white. **CDI**, N,N′-Carbonyldiimidazole.

2.5. Antibody Generation

As shown in Figure 2A, New Zealand white rabbits (6~7 weeks age) were immunized with the immunogen by subcutaneous injection to the neck and back, as described previously [29]. The immune effect was verified by detecting the titer and inhibition of the rabbit serum by the competitive indirect enzyme linked immunosorbent assay [30].

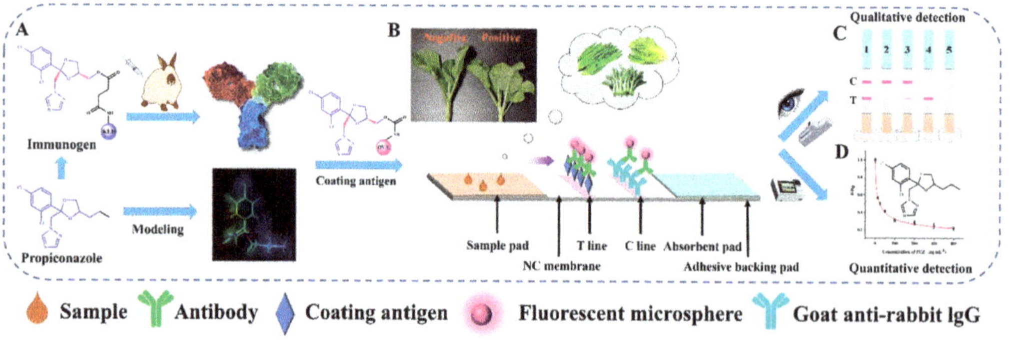

Figure 2. (**A**) Design of hapten and molecular modeling. (**B**) Schematic diagram of TRFMs-LFIA for propiconazole in brassica campestris, lettuce and romaine lettuce. (**C**) The test results from the LFIA. 1, Negative result; 2,3, positive result; 4,5, invalid results. (**D**) Quantitative detection.

The caprylic acid-ammonium sulfate precipitation method [31] was used to purify the antibody. The purified product was stored at −20 °C. Sodium dodecyl sulfate-polyacrylamide gel electrophoresis was used to check the purity of the purified antibody. The performance of antibody was preliminarily evaluated based on the 50% inhibition concentration (IC_{50}) by competitive indirect enzyme-linked immunosorbent assay [32].

2.6. Molecular Surface Electrostatic Potential Simulation

Energy-minimized three-dimensional (3D) structure and surface electrostatic potential iso-surfaces of hapten AZC-HS and 13 triazoles compounds were modelled using the Sybyl-X 2.0 program package (Tripos Inc, St. Louis, MO, USA).

2.7. Preparation of TRFM Labeled Antibody

The carboxyl group on the surface of TRFM can be linked to the amino group in the antibody by the active ester method [33]. Hence, 10 μL of TRFM was mixed with 1 mL of the MES solution (50 mM, pH 5.5). Then, 15 μL of freshly prepared EDC solution (0.5 mg/mL) and 20 μL of freshly prepared NHS solution (0.5 mg/mL) were added to the above solution. The mixture was vortexed for 10 s in order to fully disperse in solution. After activation for 15 min, the solution was centrifuged at 14,000 rpm at 4 °C for 15 min. After carefully discarding the supernatant, the white precipitate was resuspended with a boric acid buffer (BB, 20 mM, pH 8.0), and the anti-PCZ antibody (2 μL, 17.0 mg/mL) dissolved in 198 μL of BB (2 mM, pH 8.0) was added and mixed thoroughly. After incubation for another 45 min, 20 μL of 20% (*w/v*) BSA was added dropwise for 60 min of blocking reaction. The mixture was then centrifuged at 14,000 rpm at 4 °C for 15 min, and the supernatant was discarded to remove any unbound antibody and BSA. The white precipitate, which was the target TRFM-labeled antibody (TRFM-Ab) immunoconjugates, was redissolved in 200 μL phosphate buffer (PB, 20 mM, pH 7.4) containing 0.75% (*v/v*) Tween-20, 0.05% (*w/v*) NaCl, 0.5% (*w/v*) BSA, 0.3% (*w/v*) PVP, and 0.03% (*w/v*) procline-300. The resuspension was stored at 4 °C for the further use.

2.8. Fabrication of the Lateral Flow Strip

As shown in Figure 2B, the test strip mainly consists of four parts, a sample pad, a nitrocellulose membrane, an absorbent pad, and an adhesive backing pad. First, the coating antigen was diluted to 0.2 mg/mL with carbonic acid buffer and the goat anti-rabbit IgG antibody was diluted to 0.03 mg/mL with PB (20 mM, pH 7.4). Second, they were dispensed on the nitrocellulose membrane as the test (T) line and control (C) line, with 0.8 μL/cm of spray volume, respectively. The distance between the T and C line was 8 mm, then the nitrocellulose membrane was dried at 37 °C for 12 h. Third, the nitrocellulose membrane, sample pad, and absorbent pad were adhered to the adhesive backing pad as shown in Figure 2B. Finally, the assembled backing pad was cut into strips with widths of 3.05 mm and placed in a ziplock bag with silica particles as a desiccant.

2.9. Sample Preparation

Blank lettuce and romaine lettuce purchased from a local supermarket were verified as PCZ-free by HPLC analysis.

The standard addition process is to add the standard analyte of the corresponding concentration to 2 g of homogeneous vegetables, vortex, and mix for 1 min.

Samples were extracted by mixing 2 g of homogenized vegetables with 5 mL of methanol in 10 mL centrifuge tubes. After mixing and shaking for 1 min, the extracted solutions were centrifuged at 5000 rpm for 5 min. The supernatants of 4 mL were collected in a 10 mL centrifuge tube (containing 50 mg PSA, 30 mg GCB). After mixing and shaking for 1 min, the extracted solutions were centrifuged at 5000 rpm for 5 min. The supernatants of 2 mL were collected and dried with nitrogen. The substrate was reconstituted with 2 mL of PB (200 mM, pH 7.4) containing 10% (*v/v*) methanol, before being filtered through a 0.22 μm membrane and then diluted four times with PB (200 mM, pH 7.4) containing 10% (*v/v*) methanol. The final volume of 8 mL used to for TRFMs-LFIA analysis.

2.10. Principle and Detection Protocol of TRFMs-LFIA

The main principle of TRFMs-LFIA is based on the indirect competitive reaction between antibody and antigen [34]. The TRFM-Ab acts as a sensitive fluorescent probe. As shown in Figure 2B, with the help of absorption capacity of the absorbent pad, the solution can flow in the direction from the sample pad to the absorbent pad. When detecting a positive sample, the free propiconazole first reacts with the TRFM-Ab conjugates and occupies the limited TRFM-Ab binding sites. Thus, fewer or no TRFM-Ab conjugates will be captured by coating antigen on the T line. The more propiconazole exists in the sample, the weaker fluorescence response displays on the T line. In accordance with this principle, the fluorescence intensity at the test line is inversely proportional to the propiconazole concentration in the sample. If there is no fluorescence on the C line, the test strip is invalid (Figure 2C).

In this study, a vertical running mode was applied to the test strip for the detection process. In our work, 120 μL of sample solution or standard analyte was added to the microwells (microtiter plates, Guangzhou JET BIOFIL Co.), followed by mixing with 8 μL of the TRFM-Ab. After incubation for 3 min at room temperature, the test strips were vertically inserted into the microwells and incubated for another 5 min. Finally, the sample pad was discarded, and the test strips were visually inspected under ultraviolet (UV) light (Figure 2C) for qualitative results. Meanwhile, the fluorescence intensity of the T and C line were quantified by a FIC-Q1 multifunctional fluorescence reader (Figure 2D) for quantitative analysis. The ratio of T/C was used to quantify the result, where T represent fluorescence intensity at the detection line [35].

2.11. *Performance of TRFMs-LFIA*

A series of concentration of propiconazole standard solutions (0, 5, 25, 50, 75, 100, 150, and 200 ng/mL) under the optimal conditions was analyzed using the established TRFMs-LFIA to achieve the visual limit of detection (vLOD), standard curves, and quantitative limit of detection (qLOD). The vLOD, judged by the naked eyes under the UV light, is considered to be as the lowest analyte concentration that can form a significantly weaker color band on the T line than that on the control strip [36]. The standard curves were obtained by measuring a series of standard with various concentrations. Each level was tested in triplicate ($n = 3$) with the strip by FIC-Q1 fluorescence reader, which were plotted based on the B/B_0 against the propiconazole concentration, where B and B_0 were the ratio of T/C values with and without propiconazole in the sample solutions. The qLOD were defined as the concentration that gave 80% B/B_0 values according to the standard curves [37].

To evaluate the specificity of TRFMs-LFIA, 12 structural and functional analogs, including difenoconazole, diniconazole, hexaconazole, tebuconazole, epoxiconazole, myclobutanil, paclobutrazol, flusilazole, cyproconazole, triadimenol, bitertanol standard, and triadimefon, were tested along with propiconazole at the same concentration (100 ng/mL), and PB (0.1 M, 10% (*v/v*) methanol) was selected as the negative control. Each test was repeated in triplicates.

In order to confirm the reliability of the proposed TRFMs-LFIA, spiked and real-life samples were verified by HPLC-MS/MS analysis. The accuracy and precision of the developed test strip was evaluated by the recovery and coefficient of variation (CV), respectively [38]. At present, the residues of propiconazole in the brassica campestris seem to be common, and in order to better evaluate the residues of propiconazole in markets, two extra leafy vegetables, lettuce and romaine lettuce, were also selected as the test samples. Brassica campestris, lettuce, and romaine lettuce samples, including three levels of propiconazole standard concentration (10, 40, 80 μg/kg), were detected via test strip in triplicates. Blind samples (four samples of each vegetable) taken randomly from local markets in Guangzhou were tested in triplicates by both TRFMs-LFIA and HPLC-MS/MS. The quantitative consistency of TRFMs-LFIA and HPLC-MS/MS was verified by regression analysis.

3. Results and Discussion

3.1. *Hapten Design and Antibody Evaluation*

As propiconazole is too small a molecule (molar mass 342.2 g mol^{-1}) to be immunogenic and to exert an immune response to an animal body, the hapten needs to be conjugated to a macromolecular protein carrier to become a complete antigen for animal immunity [39]. However, propiconazole lacks an active group for conjugation with a carrier protein. Therefore, the structurally similar analogue molecule AZC was selected as a hapten, which extended an arm similar to propiconazole feature on its side chain. There are two chiral carbons in the molecular structure of propiconazole, and it is well known that the propiconazole contains four optical isomers. The cis-isomer generally accounts for about 60% of the racemate, and the trans-isomer accounts for about 40% [40]. Herein, AZC-HS, the cis isomer (Figure 2) was chosen as the immunizing hapten for the conjugation with carrier protein to prepare the immunogen, and then immunize rabbit to produce an antibody. It was expected to recognize the propiconazole isomer molecule, since the majority of their features are same except for a steric difference and the length of the side chain in their structures (Figure 1A).

In this conjugation, the overall structure of propiconazole was fully exposed by introducing the succinic anhydride arm, which was coupled with protein (KLH) to obtain immunogen and the conjugate curves (Figure S2A) showed the conjugate has the weak characteristic peaks of both the hapten at 260 nm and the carrier protein at 280 nm, preliminarily judging that the coupling is successful. Meanwhile, the hapten AZC was directly coupled to the OVA to obtain the coating antigen, and the conjugate curves (Figure S2B) showed that the conjugate has the characteristic peaks of both the hapten at 320 nm and the carrier protein at 280 nm, indicating successful conjugation between the hapten and the carrier (BSA or OVA) [41], thus supporting the formation of the immunogen. This immunogen was then used for antibody production.

After the rabbit antiserum was obtained, preliminary purification and evaluation were carried out per the previously reported protocol [31]. SDS-PAGE under reducing conditions (Figure S3A) shows the main protein band between 55 kDa and 180 kDa. The propiconazole antibody showed two expected protein bands around 25 kDa and 55 kDa, respectively, indicating a normal antibody feature, and consequently high antibody purity.

A competitive indirect enzyme linked immunosorbent assay was developed to preliminarily assess the usability and suitability of the obtained antibody with IC_{50}, which is the concentration resulting the half decrease in optic absorbance [32] (Figure S3B). The performance of the competitive indirect enzyme linked immunosorbent assay calibration curve exhibited a good property with IC_{50} values of 0.51 ng/mL, a working range (IC_{20}~IC_{80}) of 0.11~2.42 ng/mL, and a limit of detection (LOD) 0.06 ng/mL. According to the China food safety standard "Maximum Residue Limits of Pesticides in Foods" (GB 2763-2019), the maximum residue limits of propiconazole in other agricultural products is 0.05 mg/kg, the obtained performance of the antibody is much lower than the limit set by GB 2763-2019. This indicated that the antibody is sensitive enough to meet the limit regulation and suitable for the further investigation and assay development.

3.2. Characterization of TRFM Labeled Antibody

When TRFM was coupled with antibody, the coupling can sometimes cause a change in the particle size and zeta potential, which indicates a successful conjugation [42]. Here, the particle size is shown in Figure S3C. When the 10 μL of TRFM and TRFM-Ab was dispersed in 1 mL of ultrapure water, an average particle size of TRFM was approximately 200 nm. After coupling, its average particle size of TRFM-Ab was about 240 nm. There is a remarkable change in particle size before and after coupling. With regard to zeta potential, the potential of TRFM-Ab increased significantly from −50 mV to −35 mV of TRFMs (Figure S3D). This indicated that the antibody was successfully labeled with the TRFMs.

3.3. Optimization of TRFMs-LFIA

To obtain the best performance, each step of the TRFM-LFIA assay procedure was optimized. In this study, several key factors were optimized, such as the conditions for the preparation of TRFM-Ab (activation pH, antibody dilution buffer antibody amount, etc.)

and the assay procedure (probe amount, ion concentration of standard diluent, methanol content in diluent, coupling pH, time of coupling reaction and blocking time, etc.).

3.3.1. Activation pH

The carboxyl group on the surface of the TRFMs requires activation to couple with the amino group of the antibody [43]. Therefore, the activated pH is the key factor affecting the efficiency of activation. The activation pH was adjusted to 5.0, 5.5, 6.0, and 6.5 with MES (0.05 M) solution and other pH was adjusted to 7.0 and 7.4 with PB (0.01 M), respectively. The result showed that with the increase of pH, the fluorescence intensity of C line first increases and then decreases (Figure 3A). When the pH was at 5.5 during preparation of the TRFM labeled antibody, the fluorescence intensity of C and T lines was desired under the UV light. From the histogram analysis, the inhibition rate is the highest at the same time. Therefore, pH 5.5 was selected as the optimal activation pH.

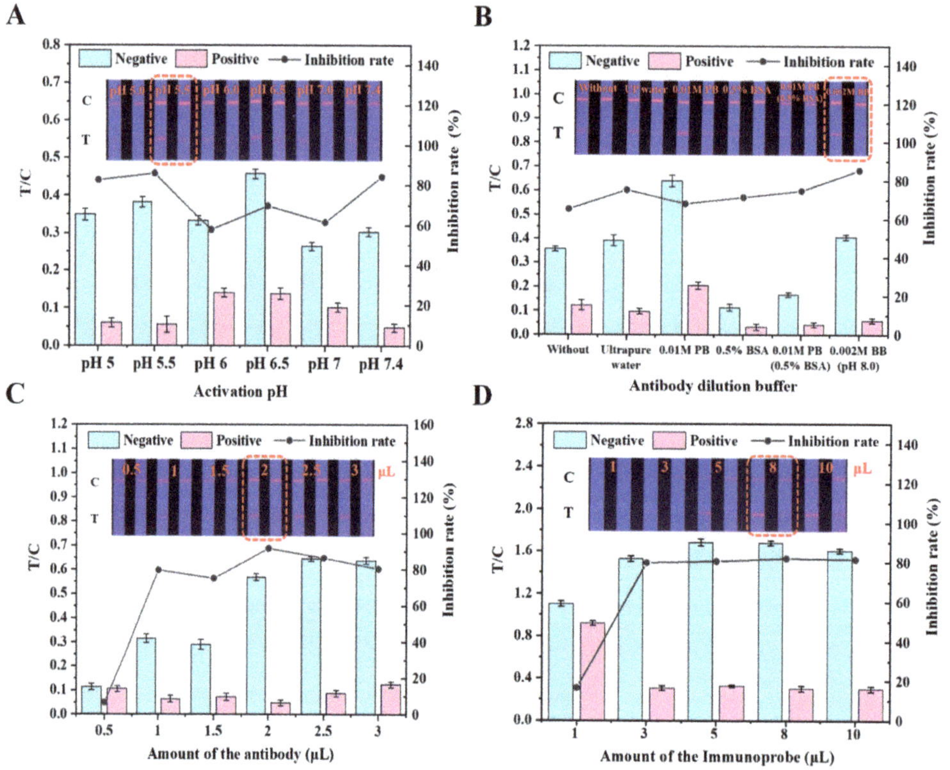

Figure 3. Optimization of working conditions. All the optimize conditions were evaluated by the negative (0 ng/mL) and positive (100 ng/mL). The values of T/C below were calculated from the pictures above by FIC-Q1 fluorescence reader and the screening criteria combined the T/C value, where B and B_0 were the ratio of T/C values with and without propiconazole in the sample solutions. Inhibition rate is equal to 1-B_0/B. (**A**) Activation pH, (**B**) Antibody dilution buffer, (**C**) Antibody amount, (**D**) Immunoprobe amount.

3.3.2. Antibody Dilution Buffer

It is known that a suitable dilution type is an important factor that can affect the sensitivity and specificity of antibody [44]. In this investigation, five antibody dilution buffers, including ultrapure water, 0.01 M PB (pH 7.4), 0.5% BSA, 0.01 M PB (0.5% BSA), and 0.002 M BB (pH 8.0), were used to dilute antibody. The results under UV light (Figure 3B) showed that the color rendering effect was the worst when the antibody dilution buffer contains 0.5% BSA, while the antibody dilution buffer contains a small number of ions and leads to a better color rendering effect. From the histogram analysis, the inhibition rate is the highest when the antibody dilution buffer was 0.002 M BB. Therefore, 0.002 M BB (pH 8.0) was selected as the optimal antibody dilution buffer.

3.3.3. Antibody Amount

To obtain an optimal antibody amount level for the assay development, five levels of antibody amount, including 0.5, 1.0, 1.5, 2.0, 2.5, and 3.0 μL (17.0 mg/mL), were used to couple with TRFM. As presented in Figure 3C, the T line fluorescence intensity gradually increased with an increase of antibody amount. When the antibody amount reaches saturation, the inhibition basically remained unchanged. From the histogram analysis, the inhibition rate is basically unchanged or even a little decreased when the antibody amount is 2 μL. Considering both the sufficient performance and cost saving, the optimal antibody amount was selected as 2 μL herein (3.4×10^{-2} mg).

3.3.4. Usage of Probe

The quantity of TRFM-Ab is directly related to the color development and cost of the test strip, the usage of TRFM-Ab was then optimized in this work. The optimization results (Figure 3D) showed that the signal intensity of the T and C lines deepened gradually with the increasing of probe volume and reached saturation when the volume was 8 μL. Thus, 8 μL of the probe was selected as the optimal usage for the further investigation.

To further improve the detection performance, ion concentration of standard diluent, methanol content in diluent, coupling pH, time of coupling reaction, and blocking time were optimized in this investigation, too (Figure S4A–E). All optimized conditions are summarized in Table 1. The screening criteria combined the T/C value, where B and B_0 are the ratio of T/C values with and without propiconazole in the sample solutions.

Table 1. Working conditions of the TRFMs-LFIA.

Working Conditions	Optimal Value
Ion concentration of standard diluent	0.2 M PB
Methanol content in diluent	10%
Activation pH	0.05 M MES (pH 5.5)
Coupling pH	0.02 M BB (pH 8.0)
Antibody dilution buffer	0.002 M BB (pH 8.0)
Antibody amount	3.4×10^{-2} mg (per strip)
Time of coupling reaction	45 min
Blocking time	60 min
Immunoprobe amount	8 μL

3.4. *Sensitivity*

According to the test procedure, a series of concentrations was selected to test with the strip. As presented in Figure 4A, when the propiconazole concentration is 100 ng/mL (green box), it can be seen that the T line intensity is significantly weaker than the control group (0 ng/mL). Therefore, 100 ng/mL was selected as the visual limit of detection (vLOD). Furthermore, a series of propiconazole standard solutions with different concentration were tested by FIC-Q1 fluorescence reader, and the obtained calibration curve (Figure 4B) showed a nonlinear fitting relationship between the B/B_0 and the propiconazole concentration with a high coefficient of determination ($R^2 = 0.987$). In addition, the qLOD was 1.92 ng/mL. The National Food Safety Standard Maximum Residue Limits of Pesticides in Foods (GB 2763-2019) sets maximum residue limits of propiconazole in other vegetables is 0.05 mg/kg. As shown in Table 2, the LOD of our proposed method was much lower than the national maximum residue and LODs using instrumental methods, demonstrating the applicability of this TRFMs-LFIA. Compared with the two reported propiconazole immunoassay methods [25,26], our method showed slightly higher LOD. Maybe the less sensitive polyclonal antibody we used was responsible for this phenomenon. However, the proposed TRFMs-LFIA not only fully meets the actual detection requirements, but also offers a sensitive as well as convenient immunoassay for propiconazole screening.

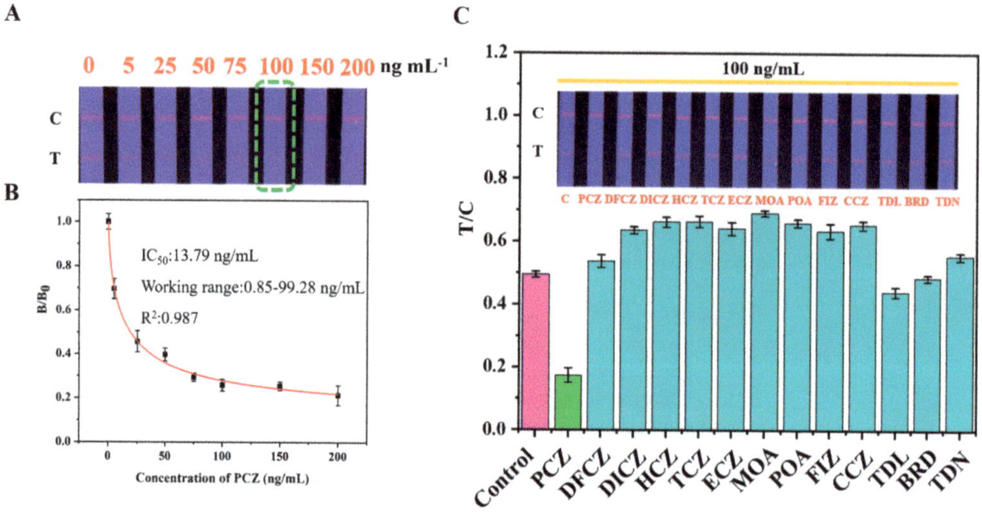

Figure 4. Assay performance of TRFMs-LFIA for propiconazole detection. (**A**) Detection results for propiconazole standard of different concentrations by the TRFMs-LFIA. Green rectangular box represents the vLOD concentrations of propiconazole by the TRFMs-LFIA. (**B**) Calibration curve of propiconazole in standard buffer. (**C**) Specificity of the TRFMs-LFIA for propiconazole detection. The concentrations of propiconazole (PCZ), difenoconazole (DFCZ), diniconazole (DICZ), hexaconazole (HCZ), tebuconazole (TCZ), epoxiconazole (ECZ), myclobutanil (MOA), paclobutrazol (POA), flusilazole (FIZ), cyproconazole (CCZ), triadimenol (TDL), bitertanol standard (BRD), and triadimefon (TDN) are all 100 ng/mL.

Table 2. Comparison of performance of different methods.

Method	Matrix	The Detection Lim (itmg/kg)	References
GC-MS	Banana	0.02	[19]
LC-MS/MS	Pepper soil	0.005 0.0015	[22]
LC-MS/MS	Soil	0.004	[21]
SPE-GC-μECD	Vegetable	0.01	[45]
GC-ECD	Groundwater	2	[15]
GC-MS	Snow peas	0.003	[20]
HPLC-MS	Soil	0.005	[46]
GC	Wolfberry	0.006	[16]
GC-ICA	Vegetable	0.00013	[25]
Ic ELISA	Vegetable	0.00026	[26]
TRFMs-LFIA	Brassica campestris	0.00192	This work

3.5. Specificity

According to the literature [47], the T/C value was chosen as the ordinate, and a histogram was used to intuitively reflect the comparison between different drugs and the blank. As shown in Figure 4C, compared with the negative control, the presence of propiconazole makes the T line disappear completely, while the other analogs did not cause obvious changes to the T line. When the buffer contains propiconazole, the T/C value of the green band is less than 0.2. The T/C value of the blue-green band with other analogs is more than twice the green band and adding other related molecules may affect the fluorescence intensity of C line, thus cause T/C value to be larger than the negative control group. This indicated that the proposed TRFMs-LFIA is highly specific for the detection of propiconazole.

The results for the 3D models (Figure 5) also confirmed the feasibility of the above results. At the lowest energy conformation, 13 triazoles and hapten AZC-HS have similar structural areas that are all in accordance with the arrangement of the 1,3-dichlorobenzene on the top, the trinitrogen ring at the bottom, and the side chain on the right. The charge distribution of hapten what we used is similar to propiconazole and the transition from benzene ring (A small amount of negative charge) to trinitrogen ring (positive charge), which implies that the obtained antibody demonstrated a good specificity. For the other 12 triazole compounds, the benzene rings of difenoconazole, diniconazole, hexaconazole, tebuconazole, paclobutrazol, flusilazole, cyproconazole, triadimenol, bitertanol standard, and triadimefon show a strong negative charge. At the same time, the positive charge of myclobutanil is mainly concentrated in the central area. Therefore, the obtained antibody cannot recognize the 12 triazole compounds, and thus demonstrated high specificity. This is also reasonable and explainable in according with the molecule modeling.

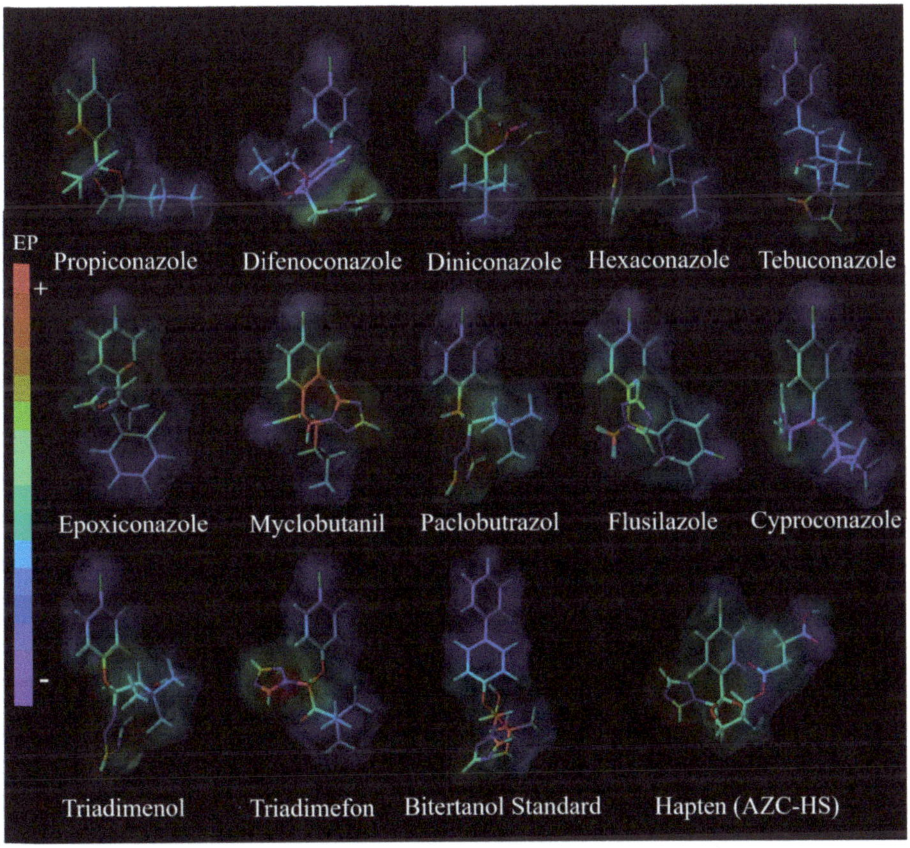

Figure 5. Lowest energy conformations and molecular electrostatic potential isosurfaces of 13 triazole compounds and hapten (Blue region represents negative charge, red region represents for positive charge. The darker the color, the stronger the charge).

3.6. Recovery

As shown in Table 3, the coefficient of variation is equal to the ratio of the standard deviation to the mean and the average recovery rate of spiked samples ranged from 78.6% to 110.7% with corresponding CVs below 16%. This indicated that it could be acceptable for a rapid screening method. The good recovery can also confirm that the chiral hapten design strategy was efficient successful, since the resultant antibody can well recognize the racemate.

Table 3. Recovery of propiconazole in brassica campestris, lettuce and romaine lettuce samples detected by TRFMs-LFIA (n = 3).

Samples	Spiked Level (ng/g)	Found ± SD (ng/g)	Recovery (%)	CV (%)
brassica campestris	10.0	8.6 ± 0.8	86.3	9.1
	40.0	44.3 ± 5.4	110.7	12.2
	80.0	78.6 ± 12.3	98.3	15.6
lettuce	10.0	8.7 ± 0.5	86.6	5.6
	40.0	36.7 ± 5.1	91.6	13.9
	80.0	83.2 ± 8.3	104.0	9.9
romaine lettuce	10.0	8.1 ± 0.7	81.2	8.8
	40.0	31.5 ± 3.7	78.6	11.8
	80.0	71.1 ± 9.7	88.9	13.6

SD, Standard deviation. CV, Coefficient of Variation.

3.7. Analysis for Blind Samples

To verify the feasibility of this method in real-life samples, brassica campestris, lettuce, and romaine lettuce were chosen and tested using the TRFMs-LFIA. The vLOD of TRFMs-LFIA in samples for propiconazole was slightly higher than that in PB, which might be attributed to the effect of the complexed food matrix. As illustrated in Figure S5, with increasing propiconazole concentration, the fluorescence intensity of T line becomes weaker under UV light. When the propiconazole concentration of brassica campestris is 150 ng/mL (red box), it can be seen that the T line intensity is significantly weaker than the control group (0 ng/mL). Similarly, 200 ng/mL was selected as the visual limit of detection (vLOD) of lettuce and romaine lettuce. Therefore, the corresponding vLOD of the brassica campestris, lettuce and romaine lettuce were 150, 200, and 200 μg/kg, respectively.

Furthermore, blind samples (four samples of each type of vegetable) were detected to evaluate the suitability of the TRFMs-LFIA for in practical applications. The data of the TRFMs-LFIA were consistent with that of the HPLC-MS/MS (Table 4). Meanwhile, a satisfactory correlation for the detection of propiconazole ($Y = 1.15X - 0.74$, $R^2 = 0.974$) was obtained with the two methods (Figure S6), indicating that the established TRFMs-LFIA had excellent reliability and could provide the point-of-care quantitative detection of propiconazole in brassica campestris, lettuce, and romaine lettuce.

To our surprise, propiconazole residues were detectable in romaine lettuce, another common vegetable on the dining table of Cantonese. It is also reported for the first time that propiconazole residue was found in romaine lettuce so far. Therefore, a comprehensive market spot check for the residue of propiconazole in romaine lettuce have to be conducted to clarify the residual status, and it is recommended that the regulatory authorities need to include the lettuce as a regulatory object.

For brassica campestris, four blind samples from the two markets all led to the detection of propiconazole residues. By calculating the dietary exposure risk entropy based on the average positive detection value (0.026 mg/kg) [11], despite the dietary exposure risk being low at present, it is necessary to monitor the propiconazole residue in vegetables, because propiconazole is not registered for use on any leafy vegetables by any country and organization.

Table 4. Comparison of propiconazole using TRFMs-LFIA and HPLC-MS/MS in blind samples (brassica campestris, lettuce and romaine lettuce) (n = 3).

Assay		TRFMs-LFIA		HPLC-MS/MS	
Samples	**Number**	**Test Value (Mean ± SD, ng/g)**	**CV (%)**	**Test Value (Mean ± SD, ng/g)**	**CV (%)**
brassica campestris	Sample 1	14.2 ± 1.6	11.5	19.0 ± 1.8	9.4
	Sample 2	64.2 ± 5.3	8.2	71.3 ± 3.5	4.9
	Sample 3	13.5 ± 1.8	13.4	9.5 ± 0.8	8.9
	Sample 4	6.6 ± 0.9	13.8	4.6 ± 0.3	7.4
lettuce	Sample 5	ND	-	ND	-
	Sample 6	ND	-	ND	-
	Sample 7	ND	-	ND	-
	Sample 8	ND	-	ND	-
romaine lettuce	Sample 9	ND	-	ND	-
	Sample 10	ND	-	ND	-
	Sample 11	ND	-	ND	-
	Sample 12	8.8 ± 0.8	9.6	12.4 ± 1.3	10.8

ND, Not detected. -, unavailable.

4. Conclusions

In conclusion, a TRFMs-LFIA for the on-site detection of propiconazole in brassica campestris, lettuce, and romaine lettuce was developed for the first time. The corresponding vLOD for propiconazole in brassica campestris, lettuce, and romaine lettuce were 150, 200, and 200 μg/kg, respectively. Blind samples were analyzed by both the TRFMs-LFIA and HPLC-MS/MS, and a good correlation between the two methods was obtained. The application of the proposed method to blind market samples, abusing risk, was also found to exist in other agricultural products besides brassica campestris for the first time. Therefore, the established method provided an idea and rapid screening tool for propiconazole monitoring and abusing risk assessment.

Supplementary Materials: The following supporting information can be downloaded at: https://www.mdpi.com/article/10.3390/foods11030324/s1, Figure S1: Ethical review of animal experiments; Figure S2: The result of hapten-carrier conjugation. (A) The UV-VIS spectroscopy of AZC-HS, KLH, and conjugates. (B) The UV-VIS spectroscopy of AZC, OVA, and conjugates. UV-VIS, Ultraviolet-visible spectroscopy. KLH, Keyhole limpet hemocyanin. OVA, Albumin from chicken egg white. CDI, N,N′-Carbonyldiimidazole; Figure S3: Characterization of PCZ Ab and TRFMs-PCZ Ab conjugation. (A) The result of SDS-PAGE. Lane M, standard protein markers. Lane 1, Before purification. Lane 2, After purification. (B) Standard curve of antibody. (C) The average particle size of TRFMs and TRFMs-PCZ Ab. (D) The zeta potential of TRFMs and TRFMs-PCZ Ab; Figure S4: Optimization of other working conditions. All the optimize conditions were evaluated by the negative (0 ng/mL) and positive (100 ng/mL). The values of T/C below was calculated from the pictures above by FIC-Q1 fluorescence reader and the screening criteria combined the T/C value, where B and B0 were the ratio of T/C values with and without propiconazole in the sample solutions. Inhibition rate is equal to 1-B0/B. (A) Standard diluent ion concentration, (B) Methanol content in diluent, (C) Coupling pH, (D) Time of coupling reaction, (E) Blocking time; Figure S5: Detection results of propiconazole in (A) brassica campestris, (B) lettuce, and (C) romaine lettuce samples by TRFM-LFIA. Red rectangular box represents the vLOD concentrations of propiconazole by TRFMs-LFIA; Figure S6: The correlation diagram of blind sample detection results of the TRFMs-LFIA and HPLC-MS/MS (n = 3).

Author Contributions: B.C.: Methodology, Investigation, Writing—original draft. X.S.: Investigation, Resources. Z.L.: Investigation, Resources. J.W.: Resources. X.L.: Validation. Z.X.: Investigation. Y.S.: Investigation. Y.L.: Investigation. X.H.: Investigation. X.W.: Investigation. H.L.: Project administration, Writing—original review & editing, Funding acquisition. All authors have read and agreed to the published version of the manuscript.

Funding: This work was financially supported by the National Key Research and Development Program of Thirteenth Five-Year Plan (No. 2017YFC1601700), the National Scientific Foundation of China (31871883, 31701703), HeYuan Planned Program in Science and Technology (210115091474673),

Generic Technique Innovation Team Construction of Modern Agriculture of Guangdong Province (2021KJ130), Lingnan Modern Agricultural Science and Technology experiment project of Guangdong Provinc (LNSYSZX001).

Data Availability Statement: Data is contained within the article (or supplementary material).

Conflicts of Interest: The authors declare no conflict of interest.

References

1. Bai, A.; Chen, A.; Chen, W.; Luo, X.; Liu, S.; Zhang, M.; Liu, Y.; Zhang, D. Study on degradation behaviour, residue distribution, and dietary risk assessment of propiconazole in celery and onion under field application. *J. Sci. Food Agric.* **2021**, *101*, 1998–2005. [CrossRef]
2. Mao, L.G.; Dong-Mei, X.U.; Tian, M.Q.; Yuan, S.K.; Fu-Gen, L.I.; Zhang, L.; Zhang, Y.N.; Jiang, H.Y. Analysis on recommended dosage of the strobilurin fungicides registered in China. *Agrochemicals* **2019**, *58*, 870–874. [CrossRef]
3. Yang, Y.J. Detailed analysis of China and global propiconazole market in 2019. *Pestic. Mark. News* **2019**, *14*, 49–51.
4. European Food Safety Authority (EFSA); Arena, M.; Auteri, D.; Barmaz, S.; Bellisai, G.; Brancato, A.; Brocca, D.; Bura, L.; Byers, H.; Chiusolo, A.; et al. Peer review of the pesticide risk assessment of the active substance propiconazole. *EFSA J.* **2017**, *15*, e04887. [CrossRef]
5. Nesnow, S.; Grindstaff, R.D.; Lambert, G.; Padgett, W.T.; Bruno, M.; Ge, Y.; Chen, P.J.; Wood, C.E.; Murphy, L. Propiconazole increases reactive oxygen species levels in mouse hepatic cells in culture and in mouse liver by a cytochrome P450 enzyme mediated process. *Chem.-Biol. Interact.* **2011**, *194*, 79–89. [CrossRef]
6. Souders, C.L., II; Xavier, P.; Perez-Rodriguez, V.; Ector, N.; Zhang, J.L.; Martyniuk, C.J. Sub-lethal effects of the triazole fungicide propiconazole on zebrafish (*Danio rerio*) development, oxidative respiration, and larval locomotor activity. *Neurotoxicology Teratol.* **2019**, *74*, 106809. [CrossRef]
7. Teng, M.; Zhao, F.; Zhou, Y.; Yan, S.; Tian, S.; Yan, J.; Meng, Z.; Bi, S.; Wang, C. Effect of propiconazole on the lipid metabolism of zebrafish embryos (*Danio rerio*). *J. Agric. Food Chem.* **2019**, *67*, 4623–4631. [CrossRef]
8. Rossi, L. Propiconazole; Pesticide tolerances. *Fed. Regist.* **2012**, *77*, 38199–38204.
9. Yang, X. The Nutritive composition and assessment of brassica parachinensis. *Food Sci. Technol.* **2002**, *9*, 74–76.
10. Xiang, D.L. Vegetable farm "dwarf medicine" to cater to the concept of deformity Zecai. *Ctry. Agric. Farmers B* **2013**, *2*, 36–37.
11. Huang, J.X.; Sun, L.; Qian, Y.E.; Gao, Y.W.; Wan, K.; Chen, H.C. Studies on effect of propiconazole on plant height and yield of flowering Chinese cabbage and its residue behavior. *China Veg.* **2019**, *3*, 47–52.
12. Qian, Y.E.; Huang, J.; Deng, Y.; Gao, Y.; Liang, Y.; Sun, L. Residue and dietary exposure risk assessment of propiconazole and chlormequat in Chinese kale, flowering Chinese cabbage and Pak-Choi in Guangzhou. *Chin. J. Trop. Crops* **2017**, *38*, 752–757.
13. Wang, C.; Wu, J.; Zhang, Y.; Wang, K.; Zhang, H. Field dissipation of trifloxystrobin and its metabolite trifloxystrobin acid in soil and apples. *Environ. Monit. Assess.* **2015**, *187*, 4100. [CrossRef] [PubMed]
14. Database, E.P. EUROPA-Plants-EU Pesticides Database. 2015. Available online: http://ec.europa.eu/sanco_pesticides/public/?event=activesubstance.selection&language=EN (accessed on 6 January 2015).
15. Song, S.L.; Zhu, R. Determination of chlorothalonil, propiconazole, hexaconazole, azoxystrobin and β-cypermethrin in groundwater by gas chromatogramy with elektron captured etector. *Rock Miner. Anal.* **2011**, *30*, 174–177. [CrossRef]
16. Yan, W.U. Gas chromatography for determination of propiconazole residues in wolfberries. *Ningxia J. Agric. For. Sci. Technol.* **2018**, *59*, 3. Available online: http://cnki.cgl.org.cn/kcms/detail/detail.aspx?&DbCode=CJFQ&filename=NXNL201801011 (accessed on 1 November 2021).
17. Mu, W.; Liu, F.; Sun, Z.; Wei, G.; Yang, L. Study on the quantitative analysis of tebuconazole, propiconazole and triadimefon by GC and HPLC. *Pestic. Sci. Adm.* **2005**, *26*, 1–3. Available online: https://xueshu.baidu.com/usercenter/paper/show?paperid=a7ecf7f87380713c5f775ca4f2b8752a&site=xueshu_se (accessed on 1 November 2021).
18. Zhang-Ming, C.; Shi-You, Y.; Rui, Z.; He, Z. Analysis of flusilazole·propiconazole by HPLC. *Agrochemicals* **2016**, *55*, 3. [CrossRef]
19. Qiong, W.U.; Wang, M.; Daizhu, L.; Xiaochun, W.U.; Huilin, G.E.; Amp, A.; Center, T. Determination and analysis of propiconazole residues in banana and soil. *Chin. J. Trop. Agric.* **2019**, *39*, 6.
20. Wang, L.Z.; Wang, D.F.; Zheng, J.C.; Wang, R.L.; Liu, Y.N.; Liang, M.; Sheng-Yu, L.U. GC-ms determination of residual fluosilazole propiconazole and difenoconazole in green pea. *Phys. Test. Chem. Anal. Part B Chem. Anal.* **2006**, *42*, 1025.
21. Blondel, A.; Krings, B.; Ducat, N.; Pigeon, O. Validation of an analytical method for 1,2,4-triazole in soil using liquid chromatography coupled to electrospray tandem mass spectrometry and monitoring of propiconazole degradation in a batch study. *J. Chromatogr. A* **2018**, *1562*, 123–127. [CrossRef]
22. Wu, S.; Zhang, H.; Zheng, K.; Meng, B.; Wang, F.; Cui, Y.; Zeng, S.; Zhang, K.; Hu, D. Simultaneous determination and method validation of difenoconazole, propiconazole and pyraclostrobin in pepper and soil by LC-MS/MS in field trial samples from three provinces, China. *Biomed. Chromatogr. Int. J. Devoted Res. Chromatogr. Methodol. Appl. Biosci.* **2018**, *32*, e4052. [CrossRef] [PubMed]
23. Preechakasedkit, P.; Pinwattana, K.; Dungchai, W.; Siangproh, W.; Chaicumpa, W.; Tongtawe, P.; Chailapakul, O. Development of a one-step immunochromatographic strip test using gold nanoparticles for the rapid detection of Salmonella typhi in human serum. *Biosens. Bioelectron.* **2012**, *31*, 562–566. [CrossRef] [PubMed]

24. Liu, L.Q.; Xu, L.G.; Suryoprabowo, S.; Song, S.S.; Kuang, H. Rapid detection of tulathromycin in pure milk and honey with an immunochromatographic test strip. *Food Agric. Immunol.* **2018**, *29*, 358–368. [CrossRef]
25. Xu, Z.K.; Meng, J.N.; Lei, Y.; Yang, X.X.; Yan, Y.Y.; Liu, H.H.; Lei, H.T.; Wang, T.C.; Shen, X.; Xu, Z.L. Highly selective monoclonal antibody-based lateral flow immunoassay for visual and sensitive determination of conazole fungicides propiconazole in vegetables. *Food Addit. Contam. Part A.* **2022**, *39*, 92–104. [CrossRef]
26. Li, J.; Ding, Y.; Chen, H.; Sun, W.; Huang, Y.; Liu, F.; Wang, M.; Hua, X. Development of an indirect competitive enzyme-linked immunosorbent assay for propiconazole based on monoclonal antibody. *Food Control* **2022**, *134*, 108751. [CrossRef]
27. Chen, L.; Sun, Y.; Hu, X.; Xing, Y.; Zhang, G. Colloidal gold-based immunochromatographic strip assay for the rapid detection of diminazene in milk. *Food Addit. Contam. Part A* **2020**, *37*, 1667–1677. [CrossRef]
28. Zhao, D.; He, L.; Pu, C.; Deng, A. A highly sensitive and specific polyclonal antibody-based enzyme-linked immunosorbent assay for detection of antibiotic olaquindox in animal feed samples. *Anal. Bioanal. Chem.* **2008**, *391*, 2653–2661. [CrossRef]
29. Liu, J.; Song, S.S.; Wu, A.H.; Kuang, H.; Liu, L.Q.; Xiao, J.; Xu, C.L. Development of immunochromatographic strips for the detection of dicofol. *Analyst* **2021**, *146*, 2240–2247. [CrossRef]
30. Fa Ng, S.; Zhang, B.; Ren, K.W.; Cao, M.M.; Shi, H.Y.; Wang, M.H. Development of a sensitive indirect competitive enzyme-linked immunosorbent assay (ic-ELISA) based on the monoclonal antibody for the detection of the imidaclothiz residue. *J. Agric. Food Chem.* **2011**, *59*, 1594–1597. [CrossRef]
31. Liang, Y.-F.; Zhou, X.-W.; Wang, F.; Shen, Y.-D.; Xiao, Z.-L.; Zhang, S.-W.; Li, Y.-J.; Wang, H. Development of a monoclonal antibody-based ELISA for the detection of *Alternaria* mycotoxin tenuazonic acid in food samples. *Food Anal. Methods* **2020**, *13*, 1594–1602. [CrossRef]
32. Lei, H.; Rui, S.; Haughey, S.A.; Qiang, W.; Xu, Z.; Yang, J.; Shen, Y.; Hong, W.; Jiang, Y.; Sun, Y. Development of a specifically enhanced enzyme-linked immunosorbent assay for the detection of melamine in milk. *Molecules* **2011**, *16*, 5591. [CrossRef]
33. Wang, D.; Zhang, Z.; Li, P.; Qi, Z.; Wen, Z. Time-resolved fluorescent immunochromatography of aflatoxin b1 in soybean sauce: A rapid and sensitive quantitative analysis. *Sensors* **2016**, *16*, 1094. [CrossRef] [PubMed]
34. Chang, X.; Zhang, Y.; Liu, H.; Tao, X. A quadruple-label time-resolved fluorescence immunochromatographic assay for simultaneous quantitative determination of three mycotoxins in grains. *Anal. Methods* **2020**, *12*, 247–254. [CrossRef]
35. Li, X.; Chen, X.; Liu, Z.; Wang, J.; Hua, Q.; Liang, J.; Shen, X.; Xu, Z.; Lei, H.; Sun, Y. Latex microsphere immunochromatography for quantitative detection of dexamethasone in milk and pork. *Food Chem.* **2021**, *345*, 128607. [CrossRef]
36. Jiang, W.; Beloglazova, N.V.; Wang, Z.; Jiang, H.; Wen, K.; Saeger, S.D.; Luo, P.; Wu, Y.; Shen, J. Development of a multiplex flow-through immunoaffinity chromatography test for the on-site screening of 14 sulfonamide and 13 quinolone residues in milk. *Biosens. Bioelectron.* **2015**, *66*, 124–128. [CrossRef] [PubMed]
37. Liu, Z.; Hua, Q.; Wang, J.; Liang, Z.; Li, X. A smartphone-based dual detection mode device integrated with two lateral flow immunoassays for multiplex mycotoxins in cereals. *Biosens. Bioelectron.* **2020**, *158*, 112178. [CrossRef]
38. Huang, D.; Ying, H.; Liu, F.; Xiaoyun, P.U. Evaluation of a time-resolved fluorescence immunochromatography for procalcitonin. *J. Third Mil. Med. Univ.* **2019**, *41*, 581–586. [CrossRef]
39. Xu, J.; Chen, Y.; Shen, P. Research progress on design and synthesis of hapten. *Jiangsu J. Agric. Sci.* **2009**, *25*, 1178–1182.
40. Geng, J.M.; Long, Y.C.; Feng, J.; Ping, L.Y. Study on the relationship between structure and activity of the sterioisomers of golden cyclotriazole fungicides. *J. Nanjing Agric. Univ.* **2003**, *26*, 102–105.
41. Gan, J.H.; Deng, W.; Jin-Ping, L.I.; Xiao-Hui, A.I. Artificial antigen synthesis and antibody preparation of doxycycline. *J. Food Sci. Biotechnol.* **2011**, *30*, 1673–1689. Available online: https://en.cnki.com.cn/Article_en/CJFDTotal-WXQG201102031.htm (accessed on 1 November 2021).
42. Miao, L.A.; Hw, B.; Js, A.; Jian, J.A.; Yy, A.; Xin, L.A.; Yz, A.; Xs, A. Rapid, on-site, and sensitive detection of aflatoxin M1 in milk products by using time-resolved fluorescence microsphere test strip. *Food Control* **2020**, *121*, 107616. [CrossRef]
43. Rodríguez-Cervantes, C.H.; Ramos, A.J.; Robledo-Marenco, M.L.; Sanchis, V.; Marín, S.; Girón-Pérez, M.I. Determination of aflatoxin and fumonisin levels through ELISA and HPLC, on tilapia feed in Nayarit, Mexico. *Food Agric. Immunol.* **2013**, *24*, 269–278. [CrossRef]
44. Li, X.; Chen, X.; Wu, X.; Wang, J.; Liu, Z.; Sun, Y.; Shen, X.; Lei, H. Rapid detection of adulteration of dehydroepiandrosterone in slimming products by competitive indirect enzyme-linked immunosorbent assay and lateral flow immunochromatography. *Food Agric. Immunol.* **2019**, *30*, 123–139. [CrossRef]
45. Ying-Jie, X.U. Determination of six triazole pesticides residues in vegetables by SPE-GC-μECD. *J. Anhui Agric. Sci.* **2014**, *42*, 5813–5815. [CrossRef]
46. Shao, Y.H. Simultaneous determination of 8 bactericide residues in soil by high performance liquid chromatography–tandem mass spectrometry. *Chem. Anal. Meterage* **2020**, *29*, 62–66. [CrossRef]
47. Zhang, H.; Wang, L.; Yao, X.; Wang, Z.; Dou, L.; Su, L.; Zhao, M.; Sun, J.; Zhang, D.; Wang, J. Developing a simple immunochromatography assay for clenbuterol with sensitivity by one-step staining. *J. Agric. Food Chem.* **2020**, *68*, 15509–15515. [CrossRef] [PubMed]

foods

MDPI

Article

Preparation of an Immunoaffinity Column Based on Bispecific Monoclonal Antibody for Aflatoxin B_1 and Ochratoxin A Detection Combined with ic-ELISA

Disha Lu, Xu Wang, Ruijue Su, Yongjian Cheng, Hong Wang, Lin Luo and Zhili Xiao *

Guangdong Provincial Key Laboratory of Food Quality and Safety, College of Food Science, South China Agricultural University, Guangzhou 510642, China; louteksa@163.com (D.L.); 18574851733@163.com (X.W.); YSQ20210721@163.com (R.S.); 18815593698@163.com (Y.C.); gzwhongd@163.com (H.W.); lin.luo@scau.edu.cn (L.L.)
* Correspondence: scau_xzl@163.com; Tel.: +86-2085280270

Abstract: A novel and efficient immunoaffinity column (IAC) based on bispecific monoclonal antibody (BsMAb) recognizing aflatoxin B_1 (AFB_1) and ochratoxin A (OTA) was prepared and applied in simultaneous extraction of AFB_1 and OTA from food samples and detection of AFB_1/OTA combined with ic-ELISA (indirect competitive ELISA). Two deficient cell lines, hypoxanthine guanine phosphoribosyl-transferase (HGPRT) deficient anti-AFB_1 hybridoma cell line and thymidine kinase (TK) deficient anti-OTA hybridoma cell line, were fused to generate a hybrid-hybridoma producing BsMAb against AFB_1 and OTA. The subtype of the BsMAb was IgG_1 via mouse antibody isotyping kit test. The purity and molecular weight of BsMAb were confirmed by SDS-PAGE method. The cross-reaction rate with AFB_2 was 37%, with AFG_1 15%, with AFM_1 48%, with AFM_2 10%, and with OTB 36%. Negligible cross-reaction was observed with other tested compounds. The affinity constant (Ka) was determined by ELISA. The Ka (AFB_1) and Ka (OTA) was 2.43×10^8 L/mol and 1.57×10^8 L/mol, respectively. Then the anti-AFB_1/OTA BsMAb was coupled with CNBr-Sepharose, and an AFB_1/OTA IAC was prepared. The coupling time and elution conditions of IAC were optimized. The coupling time was 1 h with 90% coupling rate, the eluent was methanol–water (60:40, *v:v*, pH 2.3) containing 1 mol/L NaCl, and the eluent volume was 4 mL. The column capacities of AFB_1 and OTA were 165.0 ng and 171.3 ng, respectively. After seven times of repeated use, the preservation rates of column capacity for AFB_1 and OTA were 69.3% and 68.0%, respectively. The ic-ELISA for AFB_1 and OTA were applied combined with IAC. The IC_{50} (50% inhibiting concentration) of AFB_1 was 0.027 ng/mL, the limit of detection (LOD) was 0.004 ng/mL (0.032 μg/kg), and the linear range was 0.006 ng/mL~0.119 ng/mL. The IC_{50} of OTA was 0.878 ng/mL, the LOD was 0.126 ng/mL (1.008 μg/kg), and the linear range was 0.259 ng/mL~6.178 ng/mL. Under optimum conditions, corn and wheat samples were pretreated with AFB_1-OTA IAC. The recovery rates of AFB_1 and OTA were 95.4%~105.0% with ic-ELISA, and the correlations between the detection results and LC-MS were above 0.9. The developed IAC combined with ic-ELISA is reliable and could be applied to the detection of AFB_1 and OTA in grains.

Keywords: aflatoxin B_1; ochratoxin A; bispecific monoclonal antibody; immunoaffinity column; ic-ELISA

Citation: Lu, D.; Wang, X.; Su, R.; Cheng, Y.; Wang, H.; Luo, L.; Xiao, Z. Preparation of an Immunoaffinity Column Based on Bispecific Monoclonal Antibody for Aflatoxin B_1 and Ochratoxin A Detection Combined with ic-ELISA. *Foods* **2022**, *11*, 335. https://doi.org/10.3390/foods11030335

Academic Editor: Fernando Benavente

Received: 15 December 2021
Accepted: 21 January 2022
Published: 25 January 2022

1. Introduction

Mycotoxins are natural secondary metabolites produced by filamentous fungi under suitable conditions, among which aflatoxin B_1 (AFB_1) and ochratoxin A (OTA) are the most toxic and exist widely in grains [1–4]. In 2012, aflatoxins were listed in the group 1 classification (carcinogenic to humans) by the International Agency for Research on Cancer (IARC), and OTA was classified as group 2B (possibly carcinogenic to humans) by IARC in

1993 [5]. Some research has indicated that AFB_1 and OTA frequently and simultaneously contaminated the same grain and the toxic synergistic effect caused aggravated harm [6–8].

Considering the health impact of AFB_1 and OTA, the European Union (EU) has set the maximum level of AFB_1 at 5.0 μg/kg in maize or rice, and 2.0 μg/kg in other cereals. As for OTA, the maximum level is 3.0~5.0 μg/kg in cereals [9]. In China, the maximum level for AFB_1 is mandated at 5.0~20 μg/kg in different cereals, and for OTA this is 5.0 μg/kg in all cereals [10]. Detection technology for mycotoxins in food is needed to detect and monitor mycotoxins effectively. Due to the advantages of low cost, easy operation and speed, immunoassay has become the ideal choice in large-scale screening of mycotoxins.

The preparation of antibodies in an immunoassay for mycotoxin is critical [11], and all are reported as single-specific, including polyclonal antibody [12,13], monoclonal antibody (MAb) [14–18] and recombinant antibody [19–22]. Compared with a single-specific antibody which can only recognize a single analyte in a complex food matrix, a bispecific monoclonal antibody (BsMAb) with two intrinsic specific binding sites could simultaneously recognize and bind two distinct antigens [23–25]. It is predominant in immunoassays for food safety detection [25–27], which could reduce the cost of combined use of single-specific antibodies.

Several BsMAbs have been reported and applied to establish quantitative or qualitative immunoassays in the food safety field. Wang et al. [28] developed a BsMAb-based multi-analyte enzyme-linked immunosorbent assay (ELISA) for 5-morpholinomethyl-3-amino-2-oxazolidone (AMOZ), malachite green (MG), and leuco-malachite green (LMG) detection in aquatic products. Jin et al. [29] reported a visual colloidal gold immunochromatographic strip with BsMAb to detect carbofuran and triazophos.

In addition with regards to establishing an immunoassay method, BsMAb also has significant application value in other fields of immunoassay. Heterogeneous matrix effects in food affect the detection result and lead to the reduction of sensitivity and accuracy [30,31]. This is especially the case in trace quantity analysis of mycotoxins in grains. An immunoaffinity column (IAC) could separate and concentrate mycotoxins in grain and effectively improve the matrix effect based on a specific and reversible interaction between antigen and antibody [32–34]. This may be one of the most applicable and adaptable procedures for mycotoxin detection [35]. IACs have been reported as having extracted mycotoxins in various matrices [36], such as cereals [21,34,37,38], spices [39], potables [40,41], and nuts [42]. The IACs available at present are all based on a single-specific antibody, which can only conjugate to one target compound. Simultaneous contamination of multiple mycotoxins often exists in many foods. In order to extract these mycotoxins, more IACs are needed for sample pretreatment, which would be complicated, time-consuming, and costly. Two or more different antibodies were conjugated in the same IAC in some studies [43,44], which could also recognize and extract two or more mycotoxins simultaneously. Nevertheless, more than one antibody was still necessary, and the preparation or the assay procedures were complex. Moreover, the coupling ratio of different antibodies is uncertain and hard to measure, and mutual interference may be caused by competitive binding of different antibodies. In comparison with single-specific antibody-based IAC, a BsMAb based IAC would be more efficient, convenient, and economical when the sample is contaminated with two mycotoxins. Moreover, the coupling efficiency of BsMAb based IAC is easy to measure, and the interference caused by using two or more different antibodies could be effectively avoided because of the homogeneity of the antibody structure. At present, no BsMAb based IAC has been reported. In this study, in order to further investigate the preparation and properties of BsMAb, and explore its application prospects in IAC, a BsMAb against AFB_1 and OTA was generated and characterized. Based on this BsMAb, an IAC was prepared and applied in simultaneous extraction of AFB_1 and OTA from food samples. The extraction conditions were optimized. Then the IAC was applied combined with ic-ELISA (indirect competitive ELISA) for AFB_1 and OTA detection in food samples.

2. Materials and Methods

2.1. Reagents and Material

Aflatoxin B_1 (AFB_1), ochratoxin A (OTA), aflatoxin B_2 (AFB_2), aflatoxin G_1 (AFG_1), aflatoxin G_2 (AFG_2), aflatoxin M_1 (AFM_1), aflatoxin M_2 (AFM_2), ochratoxin B (OTB), ochratoxin C (OTC), zearalenone (ZEN), deoxynivalenol (DON), T-2 toxin (T-2) and fumonisin B_1 (FB_1), N-methyl-N'-nitro-N-nitrosoguanidine (MNNG), Methyl Methane-sulfonate (MMS), 8-Azaguanine (8-AG), 5-bromo-2′-deoxyuridine (5-BrdU), Paraffin liquid, 50% PEG 4000, goat anti-mouse Immunoglobulin G horseradish peroxidase conjugate (HRP-IgG) were purchased from Sigma-Aldrich Chemical Co. (St. Louis, MO, USA). RPMI Medium 1640 basic (1×), HAT Media Supplement (50×), HT Media Supplement (50×), Pierce Rapid Isotyping Kits-Mouse were purchased from Thermo Fisher Scientific Co., Ltd. (Waltham, MA, USA). Fetal bovine serum was purchased from Guangzhou Saiguo Biotech Co., Ltd. (Guangzhou, China). All other chemicals and solvents were of analytical grade and were purchased from Shanghai Aladdin Bio-Chem Technology Co., Ltd. (Shanghai, China). Water was prepared using a Milli-Q water purification system.

CNBr-activated Sepharose 4B and polyethylene columns (8.9 mm × 63 mm) were purchased from Pharmacia Corporation (Wuhan, China). Female Balb/c mice (10 weeks old) were purchased from Guangdong Medical Laboratory Animal Centre (Foshan, China). Anti-AFB_1 hybridoma cell line E_4, anti-OTA hybridoma cell line B_3, anti-AFB_1 MAb (IgG_1), anti-OTA MAb (IgG_1), AFB_1-OVA, OTA-OVA, blocking buffer, P solution and TMB chromogenic solution were produced by our laboratory.

2.2. Apparatus

ELISA plates were washed using a Multiskan MK2 microplate washer (Thermo Fisher Scientific Inc., Waltham, MA, USA). Absorbances of ELISA were measured on a Multiskan MK3 microplate reader (Thermo Fisher Scientific Inc., Waltham, MA, USA). Absorbances of antibody concentrations were measured on a NanoDrop 2000 c Ultraviolet spectrometer (Thermo Fisher Scientific Inc., Waltham, MA, USA).The LC-MS assay was conducted using an Agilent 6400 series LC system and an ECLIPS PLUS C_{18} (2.1 × 100 mm, 1.8 μm) (Agilent Technologies, Santa Clara, CA, USA) with a Triple Quadrupole Mass Spectrometer.

2.3. Mutagenesis of HGPRT and TK Deficient Hybridoma Cell Lines

The chemical mutagen scheme was performed as follows [28,45,46]: AFB_1 hybridoma cell line E_4 was cultured in HAT medium for 2 d before mutagenic treatment in a humidified 5% CO_2 incubator at 37 °C. The cells were washed twice with PBS and cultured in HT medium for 2 d to restore the cells to normal morphology and were cultured in HAT medium for 24 h. After washing the cells with PBS twice, the cells were mutagenized in different concentrations of MNNG (5.0 μg/mL, 10.0 μg/mL, 20.0 μg/mL, and 40.0 μg/mL). The mutagenic treatment time was 2 h, 4 h, 6 h and 10 h. At the same time, the control group was cultured in 1‰ DMSO complete medium (RPMI-1640 culture media containing 20% fetal bovine serum) without MNNG. The cells were washed twice with PBS to finish the mutagenesis and cultured in complete medium. Finally, the cells were cultured for 3 d and the survival rate of the cells was determined by trypan blue staining, and the combination of MNNG concentration and mutagenic treatment time resulting in a survival rate of about 70% were selected as the optimal mutation parameters. OTA hybridoma cell line B_3 was mutated with MMS (1.0 μg/mL, 5.0 μg/mL, 10.0 μg/mL, and 20.0 μg/mL) and for 2 h, 4 h, 6 h and 10 h by the same steps as above.

2.4. Screening of HGPRT and TK Deficient Hybridoma Cell Lines

After mutagenic treatment with MNNG, AFB_1 hybridoma cells E_4 were collected and incubated in a 6-well culture plate in selective semi-solid media containing 6-TG (50 μg/mL). After 10 d, the visible white clonal cell clusters were placed into a 96-well culture plate and cultured with 200 μL complete media containing 6-TG (50 μg/mL). The cell supernatants were analyzed by ELISA to screen positive clones which could recognize and bound to AFB_1. Then the cells were tested by HAT media to confirm their HAT sensitivity. If the cell line could not grow in HAT media, it was HGPRT-deficient, and was named E_4-$HGPRT^-$. Via the same steps, the TK-deficient OTA hybridoma cell line named B_3-TK^- was screened with 5-BrdU (30 μg/mL).

2.5. Generation and Characterization of BsMAb

2.5.1. Generation of BsMAb

E_4-$HGPRT^-$ and B_3-TK^- were fused under the effect of PEG 4000 [47]. The fused cells were cultured in 96-well culture plates with HAT media for 10 d. The cell supernatants were tested by ic-ELISA to confirm the presence of BsMAb which could recognize and bound to AFB_1 and OTA simultaneously.

The selected tetra-doma cell lines were cloned by 4 rounds of limiting dilution assays, and then inoculated into female Balb/c mice that had been primed with 500 μL of sterile liquid paraffin. Ascites was collected from the Balb/c mice and purified with Protein G to obtain BsMAb. The concentration of BsMAb was measured by a NanoDrop 2000 c ultraviolet spectrometer. The subtype of BsMAb was determined by Pierce Rapid Isotyping Kits-Mouse. The purity and molecular weight of BsMAb were estimated by the SDS-PAGE method.

2.5.2. Antibody Specificity Determination

The cross-reactivity (CR) could be used as an index to evaluate the specificity of the anti-AFB_1/OTA BsMAb. Inhibition curves were fitted through ic-ELISA data to determine the IC_{50} (50% inhibiting concentration) of each common mycotoxin contaminant (aflatoxin B_1, aflatoxin B_2, aflatoxin G_1, aflatoxin G_2, aflatoxin M_1, aflatoxin M_2, ochratoxin A, ochratoxin B, ochratoxin C, zearalenone, deoxynivalenol, T-2 toxin and fumonisin B_1). The CR was calculated using the following equation:

$$CR = \frac{IC_{50} \text{ of } AFB_1 \text{ or OTA}}{IC_{50} \text{ of common mycotoxin contaminant}} \times 100\% \quad (1)$$

2.5.3. Antibody Affinity Determination

The affinity of BsMAb was validated as follows [48]: the coating antigen AFB_1-OVA and OTA-OVA diluted to four gradient concentrations (5.0 μg/mL, 2.5 μg/mL, 1.25 μg/mL, 0.625 μg/mL) were coated in microplates at 4 °C for 14 h, respectively. The anti-AFB_1/OTA BsMAb was diluted to eight gradient concentrations (9.6×10^{-3} mg/mL, 4.8×10^{-3} mg/mL, 2.4×10^{-3} mg/mL, 1.2×10^{-3} mg/mL, 6.0×10^{-4} mg/mL, 3.0×10^{-4} mg/mL, 1.5×10^{-4} mg/mL, 7.5×10^{-5} mg/mL) with PBS buffer, and then added to the micropores for specific binding. After reacting with HRP-IgG as secondary antibody, chromogenic solution, and termination solution successively, the A_{450nm} was determined. The reaction curve was fitted by Origin 8.5 software. The concentration of BsMAb was used as abscissa and A_{450nm} as ordinate. The average value of BsMAb concentration corresponding to the half-saturation index was taken as the affinity constant (Ka) of BsMAb.

2.6. Preparation of BsMAb Based IAC

According to the instruction, the procedures of IAC preparation was as follow: 0.5 g of CNBr-Sepharose 4B powder was swelled with 3 mL of HCl (1 mM) and washed with 20 mL of coupling buffer (0.1 M $NaHCO_3$, pH 8.3) 5 times. Then, the Sepharose gel was mixed with 2 mL anti-AFB_1/OTA BsMAb solution (9.61 mg in coupling buffer) and gently stirred at 25 °C. The concentration of unconjugated BsMAb in supernatant was determined by NanoDrop 2000 c ultraviolet spectrometer every 0.5 h to calculate the coupling efficiency. The optimal coupling time was determined when the coupling efficiency was over 90%.

The coupling efficiency was calculated as follows:

$$\text{Coupling Efficiency} = \frac{\text{Concentration of BsMAb before coupling} - \text{Concentration of unconjugated BsMAb}}{\text{Concentration of BsMAb Before Coupling}} \quad (2)$$

The agarose gel was redissolved in 50 mL blocking buffer (0.1 M Tris-HCl, pH 8.0) to block the free active sites at 25 °C for 2 h, and washed with 20 mL HAc-NaAc buffer (0.1 M, pH 4.0) and 20 mL Tris-HCl buffer (0.1 M Tris-HCl, pH 8.0) alternately for 4 cycles. Finally, 1 mL of immune gel prepared above was transferred to a polyethylene column and balanced with 20 mL loading buffer (0.1 M PBS, pH 7.4). The IAC prepared was stored with PBS containing 0.01% Merthiolate sodium (v/w) at 4 °C until use.

2.7. Optimization of Elution Conditions and Evaluation of the IAC Capacity

Mixed standard solution containing 50 ng AFB_1 and 50 ng OTA was diluted in 15 mL loading buffer and drawn through the IAC. Then 20 mL of loading buffer was applied to wash the IAC. Finally, elution buffer was used to elute the analyte bound to the column. The collected eluent diluted twice with PBS was detected by ic-ELISA and recovery rates were evaluated.

Three elution conditions were optimized to determine the optimum conditions, including type of elution buffer, concentration of methanol in elution buffer, and volumes of elution buffer. Elution buffer A is 0.2 M glucine solution, pH 3.0. Elution buffer B is methanol-water, 90:10, $v{:}v$. Elution buffer C is methanol-water, 60:40, $v{:}v$, pH 2.3. Elution buffer D is methanol-water, 60:40, $v{:}v$, pH 2.3, containing 1 M NaCl. The concentration of methanol in elution buffer are 30%, 40%, 50%, and 60% (v/v). The volumes of elution buffer are 1 mL, 2 mL, 3 mL, 4 mL, 5 mL.

Under the optimal conditions, 15 mL of loading buffer containing excessive mixed standard of AFB_1 (300 ng) and OTA (300 ng) was passed through the IAC. The eluent diluted with PBS was detected by ic-ELISA. The column was used repeatedly for several cycles, and the column capacity and preservation rates were calculated and compared. The column capacity/preservation rate was expressed as follows:

$$\text{Column Capacity} = \frac{\text{Concentration of } AFB_1 \text{ or OTA} \times \text{Loading Amount}}{\text{Immune Gel Volume}} \quad (3)$$

$$\text{Preservation Rate} = \frac{\text{Column Capacity for Several Cycles}}{\text{Column Capacity for the First Treatment}}. \quad (4)$$

2.8. ic-ELISA Combined with IAC

The collected eluent was detected by the ic-ELISA developed and optimized in our laboratory. The parameters of ic-ELISA are shown in Table 1 and the procedures are as follows. Coating antigen solution was coated in a 96-well polystyrene microtiter plate with 100 μL/well at 4 °C for 14 h, then washed with 300 μL/well wash solution (PBS containing 0.05% Tween-20) twice. 120 μL/well blocking buffer was used to block uncoated sites at 37 °C for 3 h, and then the plate was dried in a draught drying cabinet at 37 °C. Then, 50 μL MAb solution and 50 μL AFB_1/OTA standard solution of gradient concentrations or the collected eluent diluted twice with PBS were added to each well and incubated, and the plate was washed 4 times. 100 μL/well HPR-IgG was added for incubation, and then washed 4 times. After washing, 100 μL/well of TMB chromogenic solution was added into each well and incubated for 10 min at 37 °C, and 50 μL/well of 2 M H_2SO_4 was added to terminate the chromogenic reaction. The absorbance at 450 nm (A_{450nm}) was measured. The A_{450nm} of the well without standard solution was as B_0 and with standard solution was as B. The inhibition curve was fitted by Origin 8.5 software adopting B/B_0 versus the concentration of AFB_1 or OTA, and the IC_{50} was estimated (Figure 1). The limit of detection (LOD) was defined as the IC_{10} value. The linear range was taken as IC_{20} to IC_{80}. The IC_{50} of AFB_1 was 0.027 ng/mL, the LOD was 0.004 ng/mL (0.032 μg/kg), and the linear range was 0.006 ng/mL~0.119 ng/mL. The IC_{50} of OTA was 0.878 ng/mL, the LOD was 0.126 ng/mL (1.008 μg/kg), and the linear range was 0.259 ng/mL~6.178 ng/mL. The

LODs of the ic-ELISA to AFB_1 and OTA were obviously lower than the maximum levels. Therefore, this method was suitable for quantitative detection of AFB_1 and OTA.

Table 1. Parameters for ic-ELISA.

Target Analyte	AFB_1	OTA
Coating antigen	AFB_1-OVA	OTA-OVA
Coating concentration (μg/mL)	0.31	0.26
Coating buffer	0.05 M carbonate buffer (pH 9.6)	
Coating condition	14 h, 4 °C	
MAb	Anti-AFB_1 MAb	Anti-OTA MAb
Standard analyte	AFB_1	OTA
Competition condition	30 min, 37 °C	
HRP-IgG dilution	1:5000	
HRP-IgG incubation condition	30 min, 37 °C	

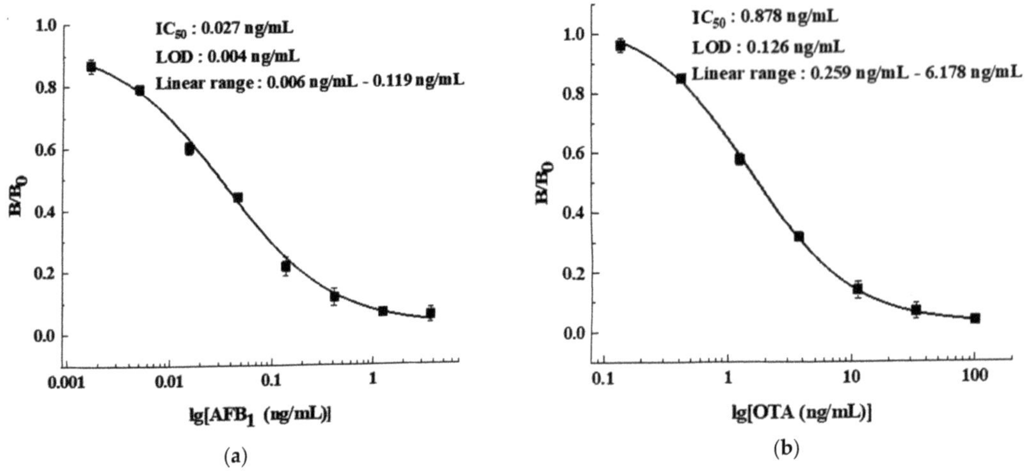

Figure 1. The inhibition curve of ic-ELISA: (**a**) AFB_1; (**b**) OTA.

2.9. Analysis of Corn and Wheat Samples

The pretreatment method for AFB_1 and OTA analysis was carried out as follows [43]. Corn and wheat samples were evenly crushed and passed through a 40-mesh sieve. 5.0 g of sample was put into a 50 mL polypropylene centrifuge tube, and 25 mL of methanol-water (70:30, *v:v*, containing 2.4% NaCl) was added, and shaken periodically for 5 min. After centrifuging at 4000 rpm for 10 min, the supernatant was collected through a 0.45 μm PTEE membrane filter to obtain the sample extract. Then 5 mL of the sample extract was diluted with 10 mL loading buffer and drawn through the AFB_1/OTA-IAC. Nonspecifically adsorbed impurities were washed off using 20 mL loading buffer. Then, elution buffer optimized in 2.7 was used to elute, and diluted twice with PBS before detection by ic-ELISA. Concurrently, the samples were detected with LC-MS.

LC-MS conditions were modified from published methods [49] as follows: an Agilent LC system coupled with an Agilent 6400 series triple Quadrupole mass spectrometer was used for the confirmatory analysis. An Agilent ECLIPS PLUS C_{18} (2.1 × 100 mm, 1.8 μm) with solvent consisted of 0.1% formic acid (mobile phase A) and acetonitrile (mobile phase B, 0–2 min, 30%; 2–5 min, 90%; 5–10 min, 5%) at a flow rate of 0.30 mL/min. The temperature was 40 °C and the injection volume was 5 μL. The analysis was performed using positive-ion electrospray interface (ESI) with a multiple reaction monitoring (MRM) mode. The retention times of AFB_1 and OTA were 3.339 min and 3.94 min, respectively.

Blank corn and wheat samples without AFB_1 and OTA, having been confirmed by LC-MS, were used to spike recovery experiments. AFB_1 and OTA of different concentrations (4 μg/kg, 10 μg/kg, 20 μg/kg, 50 μg/kg, 100 μg/kg) were spiked into the blank samples. Eight grain samples were randomly selected from a local market, including four corn samples and four wheat samples. After pretreatment, the sample eluents were detected by ic-ELISA, and the results were compared with those of LC-MS.

3. Results

3.1. Mutagenesis and Screening of HGPRT and TK Deficient Hybridoma Cell Lines

The AFB_1 hybridoma cell line E_4 and OTA hybridoma cell line B_3 were applied for construction of a HGPRT and TK deficient mutant, respectively. When the concentration of MNNG was 10 μg/mL and the treatment time was 6 h, the cell survival percentage was 69.9% (Figure 2a), which was determined as the optimal mutation condition of E_4 cells. When the concentration of MMS was 5 μg/mL and the treatment time was 4 h, the cell survival percentage was 70.2% (Figure 2b), and the optimal mutation condition for B_3 cells was determined. As mutagens, MNNG and MMS have strong mutagenic effect on cells. Over-high concentration and overlong treating time will reduce the activity of cells, or even result in apoptosis.

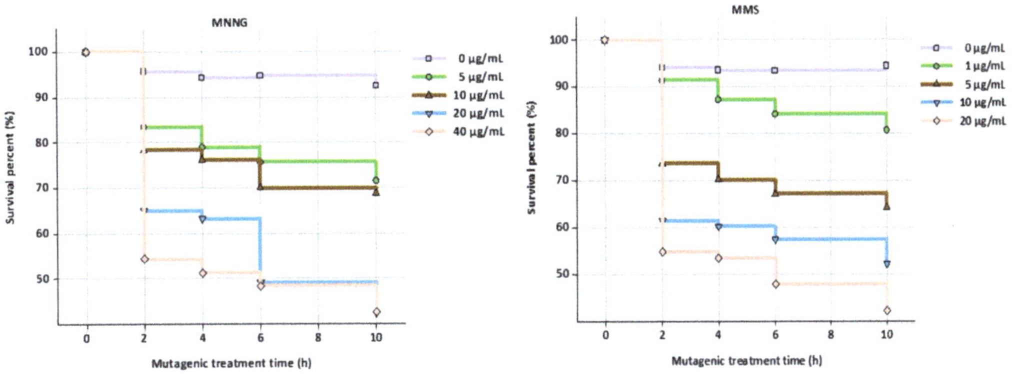

(a) (b)

Figure 2. Effects of mutagen concentration and mutagenic treatment time on survival percentage (%) of the hybridoma cell line. (**a**) AFB_1 hybridoma cell line E_4; (**b**) OTA hybridoma cell line B_3.

After MNNG mutation treatment, the E_4 cells were inoculated in 6-TG medium for selective screening. The screened E_4-HGPRT$^-$ cell line gradually formed colonies with the ability to fission and grow. Finally, stable inherited deficient cells E_4-HGPRT$^-$ were obtained. Similarly, stable inherited deficient cells B_3-TK$^-$ were screened in 5-BrdU. Figure 3 depict the growth curves of E_4-HGPRT$^-$ and B_3-TK$^-$ in screening reagent (6-TG and 5-BrdU), complete media and HAT media after mutagenic treatment. The unmutated cells could not survive in screening reagent (Figure 3a,d). In complete medium culture, both normal cell line and mutant could grow normally (Figure 3b,e). The cell apoptosis in HAT medium was adopted as an indication of HGPRT or TK deficiency. The growth curve (Figure 3c,f) shows that E_4-HGPRT$^-$ and B_3-TK$^-$ were unable to divide in HAT media, which indicates that E_4-HGPRT$^-$ and B_3-TK$^-$ were sensitive to HAT media.

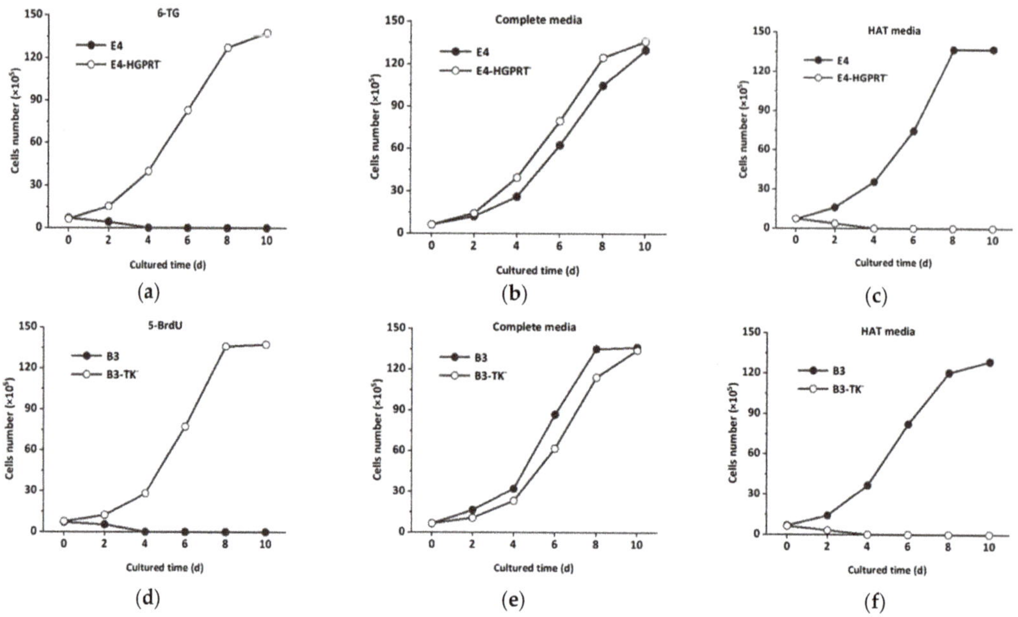

Figure 3. The growth curve of AFB_1 hybridoma cell lines E_4 and OTA hybridoma cell lines B_3: (**a**) cell line E_4 and E_4-HGPRT$^-$ in 6-TG; (**b**) cell line E_4 and E_4-HGPRT$^-$ in the complete media; (**c**) cell line E_4 and E_4-HGPRT$^-$ in HAT media; (**d**) cell line B_3 and B_3-TK$^-$ in 5-BrdU; (**e**) cell line B_3 and B_3-TK$^-$ in the complete media; (**f**) cell line B_3 and B_3-TK$^-$ in HAT media.

3.2. Generation and Characterization of BsMAb

3.2.1. Generation of BsMAb

E_4-HGPRT$^-$ and B_3-TK$^-$ were fused by hybrid-hybridoma technology. A total of 83 tetradoma colonies were generated, and the cell supernatant was identified by ELISA. After 3–4 rounds of subcloning, a positive tetradoma T26 was chosen and was used for production of anti-AFB_1/OTA BsMAb by induction in vivo. Clear bands of heavy chain (about 55 kDa) and light chain (about 25 kDa) from BsMAb were observed in the electropherogram (Figure 4). The concentration of BsMAb was 9.61 mg/mL measured by NanoDrop ultraviolet spectrophotometer.

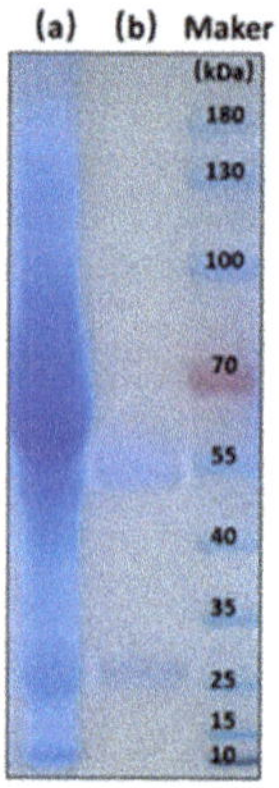

Figure 4. The electropherogram of BsMAb: (**a**) ascites; (**b**) purified BsMAb.

3.2.2. Antibody Specificity Determination

The cross-reactivity of the BsMAb with other common mycotoxins was detected by ELISA. As shown in Table 2, five mycotoxins containing similar structures displayed evident cross-reactivities. The cross-reaction rate with AFB_2 was 37%, with AFG_1 15%, with AFM_1 48%, with AFM_2 10%, and with OTB 36%. Negligible cross-reaction was observed with other tested compounds.

Table 2. IC_{50} and cross-reactivity (CR) of anti-AFB_1/OTA BsMAb against related mycotoxins.

Mycotoxin Analyte	Structure	IC_{50} (ng/mL)	CR (%)
AFB_1	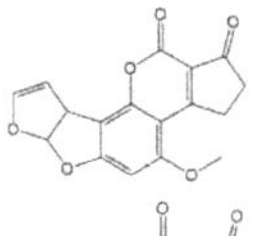	0.037	100
AFB_2	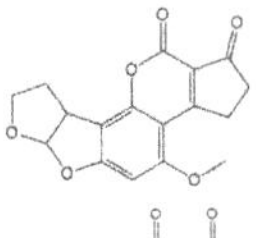	0.101	37
AFG_1	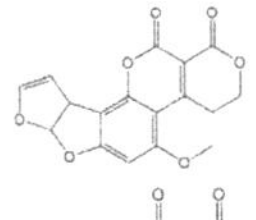	0.252	15
AFG_2	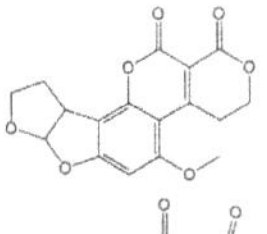	3.584	1
AFM_1	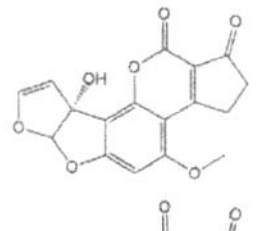	0.077	48
AFM_2	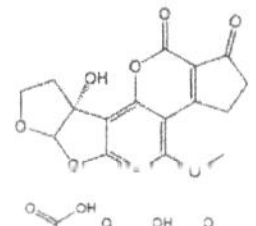	0.377	10
OTA	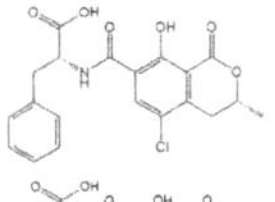	2.040	100
OTB	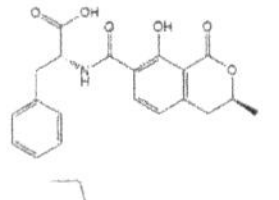	5.620	36
OTC	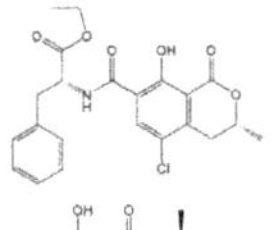	201.535	1
ZEN	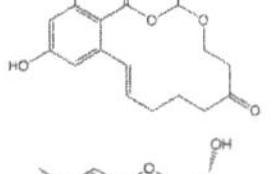	>1000	<0.1
DON	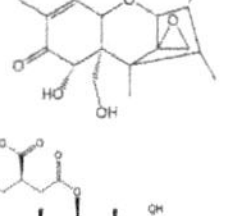	>1000	<0.1
FB_1	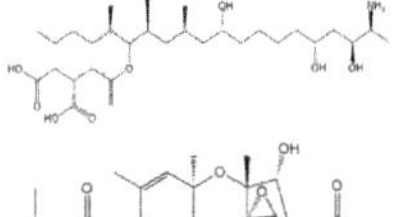	>1000	<0.1
T-2	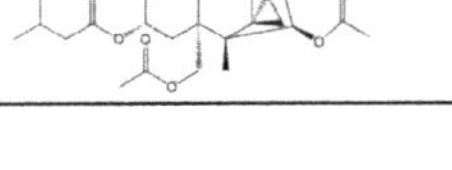	>1000	<0.1

It may be speculated from the cross-reactivities of aflatoxins that the specific recognition site of the BsMAb to aflatoxins was mainly coumarin plus cyclopentenone, followed by the difuran ring. The double bond on the furan ring was more conducive to the recognition of the antibody than the single bond (the cross-reaction rate of AFG_1 was higher than that of AFG_2, and the same trend exists between AFM_1 and AFM_2). Very often anti-AFB_1 antibodies have high cross-reactivity for other aflatoxins, and a positive result for AFB_1 is a sufficient motive for detailed analysis of aflatoxins in the sample by other techniques [43]. IAC was applied as pretreatment method in this study, and a certain amount of cross-reactivity of the antibody would be more acceptable than that used in specific quantitative analysis.

3.2.3. Antibody Affinity Determination

The affinity constant (Ka) at different coating concentrations was confirmed by ELISA and shown in Figure 5. The average Ka of the BsMAb to AFB_1 and OTA were 2.43×10^8 L/mol, and 1.57×10^8 L/mol, respectively. Previous studies indicated that the affinity constant (Ka) of high affinity antibodies is between 10^7 L/mol–10^{12} L/mol [48], suggesting that the BsMAb prepared in this study is a high affinity antibody.

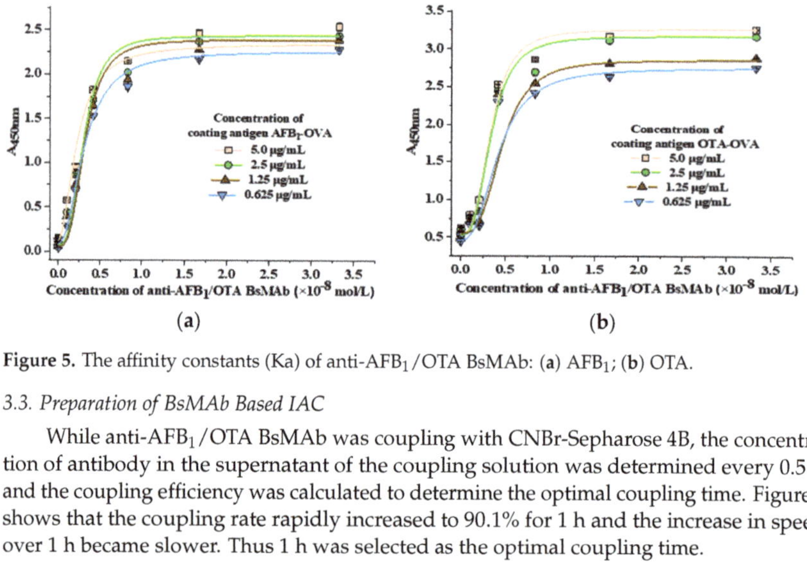

Figure 5. The affinity constants (Ka) of anti-AFB_1/OTA BsMAb: (**a**) AFB_1; (**b**) OTA.

3.3. Preparation of BsMAb Based IAC

While anti-AFB_1/OTA BsMAb was coupling with CNBr-Sepharose 4B, the concentration of antibody in the supernatant of the coupling solution was determined every 0.5 h, and the coupling efficiency was calculated to determine the optimal coupling time. Figure 6 shows that the coupling rate rapidly increased to 90.1% for 1 h and the increase in speed over 1 h became slower. Thus 1 h was selected as the optimal coupling time.

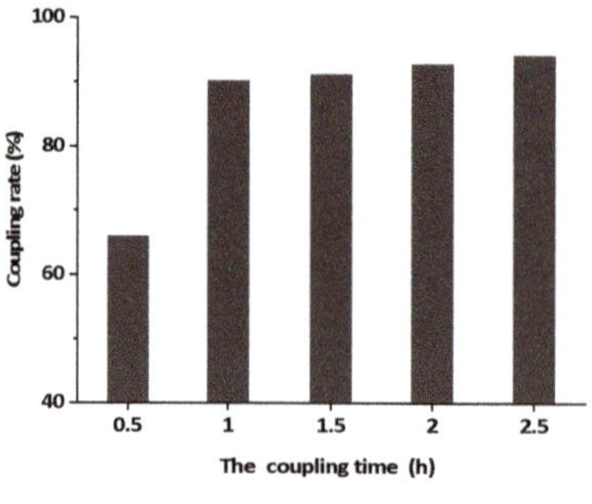

Figure 6. The coupling rate of different coupling times.

3.4. Optimization of Elution Conditions and Evaluating of the IAC Capacity

To achieve high extraction efficiency, three affecting conditions were optimized and were evaluated by the recovery of AFB_1 and OTA.

Firstly, four kinds of elution buffer (A, B, C, D) were tested. The recoveries are shown in Figure 7a. Solution A achieved the lowest recovery, containing no methanol and ionic compound. When the column was eluted with solution D, the recovery of AFB_1 and OTA was up to 90%. The concentration of methanol, pH and ion concentration could all influence the recovery. The methanol concentration has significant effect, and ion concentration and low pH also improved the elution recovery. It can be seen from Figure 7a that the recoveries of solution B and D are all around 80%. However, high methanol concentration may affect the activity of the antibody [43], so solution D containing NaCl and medium concentration of methanol is preferred.

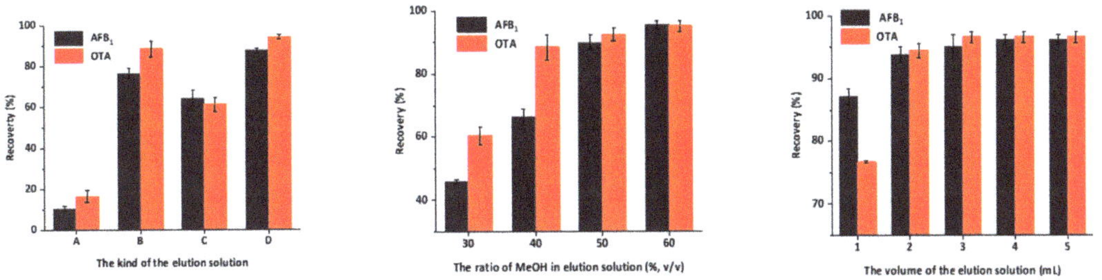

(a) (b) (c)

Figure 7. Optimization results of the IAC elution conditions: (**a**) the recovery of AFB_1 and OTA with different kind of elution solution ($n = 3$); (**b**) the recovery of AFB_1 and OTA with different ratios of methanol–water as elution solution ($n = 3$); (**c**) The recovery of AFB_1 and OTA with different elution volume ($n = 3$).

Secondly, to further investigate the effect of methanol concentration, elution solutions containing different concentrations of methanol (30%, 40%, 50% and 60%, v/v, 1 M NaCl, pH 2.3) were evaluated. The recovery rose along with the increase of methanol concentration, and the highest recovery was obtained when the concentration of methanol was 60% (Figure 7b).

Thirdly, under the conditions optimized above, using solution containing 60% methanol, 1 M NaCl, pH 2.3 as elution solution, the effect of elution solution volumes (1, 2, 3, 4, 5 mL) on recoveries were also investigated. As shown in Figure 7c, recoveries increased from 87.2% to 96.2% for AFB_1 and 76.7% to 96.6% for OTA as the volume increased from 1 mL to 4 mL. The highest recovery was acquired when the volume was above 4 mL, and further increase of the volume did not visibly improve the recovery, thus 4 mL was selected as the optimized volume.

Under the optimized elution conditions (4 mL of solution D with 60% methanol and 1 M NaCl used as elution buffer), the IAC was used repeatedly for seven cycles and the capacity was detected at every cycle. As shown in Table 3, the maximum binding capacities of the IAC for AFB_1 and OTA were both over 165 ng. As the column is used repeatedly, the capacity decreases gradually. After seven cycles of use, the preservation rates of column capacity for AFB_1 and OTA were 69.3% and 68.0%, respectively (Table 3), which is equivalent to 114.4 μg/kg and 116.3 μg/kg of AFB_1 and OTA in samples, respectively, by conversion. According to some investigation results of mycotoxin contamination in agro–products and food samples [50–53], the capacity of the IAC could meet the needs of practical applications.

Table 3. Column capacity and preservation rate of AFB_1 and OTA in 7 cycles (n = 3).

Cycle	AFB_1		OTA	
	Column Capacity (ng)	Preservation Rate (%)	Column Capacity (ng)	Preservation Rate (%)
1	165.0	100.0	171.1	100.0
2	162.5	98.5	169.5	99.1
3	159.0	96.4	166.4	97.3
4	153.6	93.1	156.2	91.3
5	147.5	89.4	145.8	85.2
6	128.9	78.1	133.3	77.9
7	114.4	69.3	116.3	68.0

3.5. Analysis of Corn and Wheat Samples

AFB_1 and OTA in spiked corn and wheat samples were detected by IAC combined with ELISA, and confirmed by LC-MS. The recoveries of IAC-ELISA ranged from 95.4% to 105.0%, and the coefficients of variation (CV) were less than 10% (Table 4). The results indicated that the assay performed well with appropriate recovery and accuracy. This may be ascribed to IAC pretreatment, which could relieve matrix interference effectively and ameliorate the recovery and accuracy of ELISA for AFB_1 and OTA detection in food matrices. The results were confirmed by LC-MS to further prove the reliability of this assay.

Table 4. Results of spiked recovery experiments of corn and wheat samples with IAC-ELISA and LC-MS (n = 2).

Samples	Analyte	Spiked Concentration (μg/kg)	IAC-ELISA			LC-MS		
			Measured (μg/kg)	Recovery (%)	CV (%)	Measured (μg/kg)	Recovery (%)	CV (%)
Corn 0	AFB_1	0	ND [a]	NC [b]	NC	ND	NC	NC
	OTA	0	ND	NC	NC	ND	NC	NC
Corn 1	AFB_1	100	95.6	95.6	5.0	97.3	97.3	3.3
	OTA	100	98.6	98.6	4.5	97.9	97.9	3.5
Corn 2	AFB_1	50	49.2	98.4	4.7	49.8	99.6	3.3
	OTA	50	49.1	98.2	4.4	49.4	98.8	3.1
Corn 3	AFB_1	20	19.2	96.0	4.3	19.2	96.0	3.5
	OTA	20	19.6	98.0	4.4	19.2	96.0	3.2
Corn 4	AFB_1	10	9.9	99.0	3.4	9.8	98.0	3.8
	OTA	10	10.3	103.0	4.2	9.8	98.0	3.6
Corn 5	AFB_1	4	4.1	102.5	2.6	3.8	95.0	3.2
	OTA	4	4.2	105.0	2.4	3.9	97.5	3.5
Wheat 0	AFB_1	0	ND	NC	NC	ND	NC	NC
	OTA	0	ND	NC	NC	ND	NC	NC
Wheat 1	AFB_1	100	99.1	99.1	4.8	98.2	98.2	3.1
	OTA	100	98.9	98.9	4.6	97.5	97.5	3.1
Wheat 2	AFB_1	50	47.7	95.4	4.5	48.7	97.4	3.8
	OTA	50	47.7	95.4	3.3	49.2	98.4	3.5
Wheat 3	AFB_1	20	19.5	97.5	4.0	19.2	96.0	3.2
	OTA	20	19.4	97.0	3.3	19.7	98.5	3.1
Wheat 4	AFB_1	10	9.6	96.0	3.3	9.8	98.0	3.8
	OTA	10	9.9	99.0	3.2	9.9	99.0	3.5
Wheat 5	AFB_1	4	3.9	97.5	3.1	3.9	97.5	3.2
	OTA	4	3.9	97.5	2.8	4.0	100.0	3.8

[a] ND, not detectable. [b] NC, not calculated.

Eight real samples of corn and wheat were simultaneously analyzed by the established IAC-ELISA and LC-MS. The results are shown in Figure 8. AFB_1 and OTA both tested positive in the eight samples, but the levels were all below the maximum levels. The correlations between the detection results and LC-MS were both above 0.9. Thus it can be seen that the established IAC-ELISA could be applied to the detection of AFB_1 and OTA simultaneously in real corn and wheat samples with high accuracy.

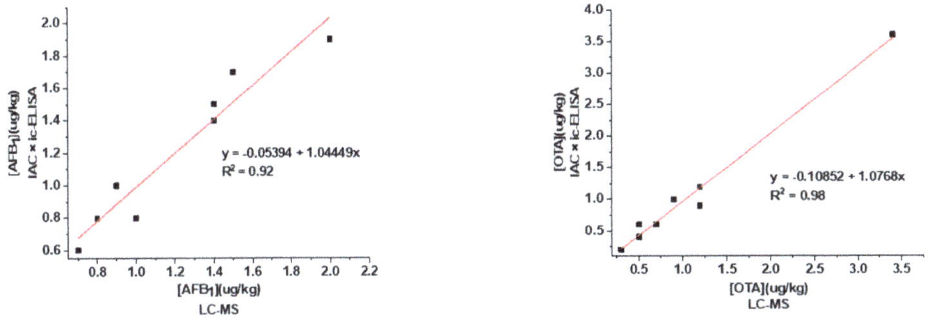

(a) (b)

Figure 8. Results of real sample detection by IAC-ELISA and LC-MS: (**a**) AFB_1; (**b**) OTA.

4. Conclusions

In this study, we develop a BsMAb based IAC which could bind AFB_1 and OTA simultaneously, and the binding sites were 1:1 with the two mycotoxins. The coupling rate and the binding site ratio of AFB_1 and OTA are easy to adjust and measure, and mutual interference could be avoided because of the homogeneity of the antibody structure. It is more efficient, convenient, and economical than the IACs based on single-specific antibodies. With satisfactory matrix effect elimination effect and recovery rate, it could be applied to detect AFB_1 and OTA rapidly and effectively, combined with ic-ELISA. Besides, the application of this anti-AFB_1/OTA BsMAb in IAC could provide reference for the application of other BsMAb in IAC in the future.

The BsMAb prepared in this study demonstrated different cross-reactivities to five biotoxins with similar structures. It is necessary to further improve the hapten structures and the screening strategy to prepare a more specific BsMAb if it were to be applied in the field with higher requirements for specificity. As a mature ELISA detection mode that could quantitatively determine two analytes simultaneously is unavailable at present, the ic-ELISA used in this study can just detect AFB_1 and OTA respectively. If simultaneous detection of two toxins in ELISA can be achieved, this study will have more practical application value.

Author Contributions: Conceptualization, D.L.; Data curation, D.L., X.W. and R.S.; Formal analysis, Y.C. and L.L.; Supervision, Z.X.; Writing—original draft, D.L.; Writing—review & editing, H.W. and Z.X. All authors have read and agreed to the published version of the manuscript.

Funding: This study was supported financially by the National Key Research and Development Program of China (2018YFC1602903) and Guangdong Basic and Applied Basic Research Foundation (2019A1515012107).

Institutional Review Board Statement: All procedures involving animals were performed in accordance with the protective and administrant laws for laboratory animals of China and approved by the Institutional Authority for Laboratory Animal Care.

Informed Consent Statement: Not applicable.

Data Availability Statement: Data is contained within the article.

Conflicts of Interest: All authors declare no conflict of Interest.

References

1. Duarte, S.C.; Lino, C.M.; Pena, A. Ochratoxin A in feed of food-producing animals: An undesirable mycotoxin with health and performance effects. *Vet. Microbiol.* **2011**, *154*, 1–13. [CrossRef] [PubMed]
2. Egmond, H.; Dekker, W.H. Worldwide regulations for mycotoxins. *Nat. Toxins* **2010**, *3*, 332–336. [CrossRef] [PubMed]
3. Fuchs, R.; Peraica, M. Ochratoxin A in human kidney diseases. *Food Addit. Contam.* **2005**, *22*, 53–57. [CrossRef] [PubMed]
4. Santacroce, M.P.; Conversano, M.C.; Casalino, E.; Lai, O.; Zizzadoro, C.; Centoducati, G.; Crescenzo, G. Aflatoxins in aquatic species: Metabolism, toxicity and perspectives. *Rev. Fish Biol. Fish.* **2008**, *18*, 99–130. [CrossRef]
5. WHO International Agency for Research on Cancer. IARC Monographs on the Identification of Carcinogenic Hazards to Humans. 2021. Available online: https://monographs.iarc.who.int/list-of-classifications (accessed on 14 December 2021).
6. Kumar, A.; Jindal, N.; Shukla, C.L.; Pal, Y.; Ledoux, D.R.; Rottinghaus, G.E. Effect of ochratoxin A on Escherichia coli-challenged broiler chicks. *Avian Dis.* **2003**, *47*, 15–424. [CrossRef]
7. Alassane-Kpembi, I.; Schatzmayr, G.; Taranu, I.; Marin, D.; Puel, O.; Oswald, I.P. Mycotoxins co-contamination: Methodological aspects and biological relevance of combined toxicity studies. *Crit. Rev. Food Sci. Nutr.* **2016**, *57*, 3489–3507. [CrossRef]
8. Madrigal-Santillan, E.; Morales-Gonzalez, J.A.; Vargas-Mendoza, N.; Reyes-Ramírez, P.; Cruz-Jaime, S.; Sumaya-Martínez, T.; Pérez-Pastén, R.; Madrigal-Bujaidar, E. Antigenotoxic Studies of Different Substances to Reduce the DNA Damage Induced by Aflatoxin B-1 and Ochratoxin A. *Toxins* **2010**, *2*, 738–757. [CrossRef]
9. European Commission. Commission Regulation (EC) No. 1881/2006 of 19 December 2006 setting maximum levels for certain contaminants in foodstuffs. *Off. J. Eur. Union* **2006**, *L364*, 20–23.
10. *GB2761-2017*; Chinese National Food Safety Standard, Maximum Residue Limits for Mycotoxins in Food. National Medical Products Administration: Beijing, China, 2017.
11. Amino, N.; Hidaka, Y. Various types of immunoassay. *Nihon Rinsho Jpn. J. Clin. Med.* **1995**, *53*, 2107–2111.
12. Lee, N.; Wang, S.; Allan, R.D.; Kennedy, I.R. A Rapid Aflatoxin B_1 ELISA: Development and Validation with Reduced Matrix Effects for Peanuts, Corn, Pistachio, and Soybeans. *J. Agric. Food. Chem.* **2004**, *52*, 2746–2755. [CrossRef] [PubMed]
13. Sun, X.; Zhao, X.; Tang, J.; Jun, Z.; Chu, F.S. Preparation of gold-labeled antibody probe and its use in immunochromatography assay for detection of aflatoxin B_1. *Int. J. Food Microbiol.* **2005**, *99*, 185–194.
14. Huang, X.; Huang, T.; Li, X.; Huang, Z. Flower-like gold nanoparticles-based immunochromatographic test strip for rapid simultaneous detection of fumonisin B_1 and deoxynivalenol in Chinese traditional medicine. *J. Pharm. Biomed. Anal.* **2019**, *177*, 112895. [CrossRef] [PubMed]
15. Machado, J.M.; Soares, R.R.; Chu, V.; Conde, J.P. Multiplexed capillary microfluidic immunoassay with smartphone data acquisition for parallel mycotoxin detection. *Biosens. Bioelectron.* **2018**, *99*, 40–46. [CrossRef] [PubMed]
16. Gong, Y.; Zheng, Y.; Jin, B.; You, M.; Wang, J.; Li, X.; Lin, M.; Xu, F.; Li, F. A portable and universal upconversion nanoparticle-based lateral flow assay platform for point-of-care testing. *Talanta* **2019**, *201*, 126–133. [CrossRef] [PubMed]
17. Qu, J.; Xie, H.; Zhang, S.; Luo, P.; Guo, P.; Chen, X.; Ke, Y.; Zhuang, J.; Zhou, F.; Jiang, W. Multiplex Flow Cytometric Immunoassays for High-Throughput Screening of Multiple Mycotoxin Residues in Milk. *Food Anal. Methods* **2019**, *12*, 877–886. [CrossRef]
18. Qileng, A.; Liang, H.; Huang, S.; Liu, W.; Xu, Z.; Liu, Y. Dual-function of ZnS/Ag 2 S nanocages in ratiometric immunosensors for the discriminant analysis of ochratoxins: Photoelectrochemistry and electrochemistry. *Sens. Actuators B Chem.* **2020**, *314*, 128066. [CrossRef]
19. Han, Z.; Tang, Z.; Jiang, K.; Huang, Q.; Meng, J.; Nie, D.; Zhao, Z. Dual-target electrochemical aptasensor based on co-reduced molybdenum disulfide and Au NPs (rMoS 2 -Au) for multiplex detection of mycotoxins. *Biosens. Bioelectron.* **2020**, *150*, 111894. [CrossRef]
20. Shen, Y.; Pan, D.; Li, G.; Hu, H.; Xue, H.; Zhang, M.; Zhu, M.; Gong, X.; Zhang, Y.; Wan, Y. Direct Immunoassay for Facile and Sensitive Detection of Small Molecule Aflatoxin B_1 based on Nanobody. *Chemistry* **2018**, *24*, 9869–9876.
21. Liu, H.; Zhang, J.; Ding, K.; Chen, X.; Han, T. The development and characterization of an immunoaffinity column used for the simultaneous selective extraction of Fusarium toxins from grain products. *Qual. Assur. Saf. Crops Foods* **2019**, *11*, 325–331. [CrossRef]
22. Wang, F.; Li, Z.; Yang, Y.; Wan, D.; Vasylieva, N.; Zhang, Y.; Cai, J.; Wang, H.; Shen, Y.; Xu, Z. Chemiluminescent Enzyme Immunoassay and Bioluminescent Enzyme Immunoassay for Tenuazonic Acid Mycotoxin by Exploitation of Nanobody and Nanobody-Nanoluciferase Fusion. *Anal. Chem.* **2020**, *72*, 11935–11942. [CrossRef]
23. Moldenhauer, G. Bispecific Antibodies from Hybrid Hybridoma. In *Bispecific Antibodies*; Kontermann, R., Ed.; Academic Press: Cambridge, MA, USA; Springer: Berlin/Heidelberg, Germany, 2011; pp. 29–46.
24. Sedykh, S.E.; Prinz, V.V.; Buneva, V.N.; Nevinsky, G.A. Bispecific antibodies: Design, therapy, perspectives. *Drug Des. Dev. Ther.* **2018**, *12*, 195. [CrossRef] [PubMed]
25. Byrne, H.; Conroy, P.J.; Whisstock, J.C.; O'Kennedy, R.J. A tale of two specificities: Bispecific antibodies for therapeutic and diagnostic applications. *Trends Biotechnol.* **2013**, *31*, 621–632. [CrossRef] [PubMed]
26. Ouyang, H.; Wang, L.; Yang, S.; Wang, W.; Wang, L.; Liu, F.; Fu, Z. Chemiluminescence reaction kinetics-resolved multianalyte immunoassay strategy using a bispecific monoclonal antibody as the unique recognition reagent. *Anal. Chem.* **2015**, *87*, 2952. [CrossRef] [PubMed]
27. Takahashi, M.; Fuller, S.A. Production of murine hybrid-hybridomas secreting bispecific monoclonal antibodies for use in urease-based immunoassays. *Clin. Chem.* **1988**, *34*, 1693–1696. [CrossRef] [PubMed]

28. Wang, F.; Wang, H.; Shen, Y.; Li, Y.; Dong, J.; Xu, Z.; Yang, J.; Sun, Y.; Xiao, Z. Bispecific Monoclonal Antibody-Based Multianalyte ELISA for Furaltadone Metabolite, Malachite Green, and Leucomalachite Green in Aquatic Products. *J. Agric. Food. Chem.* **2016**, *64*, 8054–8061. [CrossRef] [PubMed]
29. Jin, R.; Guo, Y.; Wang, C.; Wu, J.; Zhu, G. Development of a Bispecific Monoclonal Antibody to Pesticide Carbofuran and Triazophos Using Hybrid Hybridomas. *J. Food Sci.* **2009**, *74*, T1–T6. [CrossRef]
30. Trucksess, M.W.; Weaver, C.M.; Oles, C.J.; Oles, C.J.; Fry, F.S.; Noonan, G.O.; Betz, J.M.; Rader, J.I. Determination of aflatoxins B_1, B_2, G_1, and G_2 and ochratoxin A in ginseng and ginger by multitoxin immunoaffinity column cleanup and liquid chromatographic quantitation: Collaborative study. *J. AOAC Int.* **2008**, *91*, 511–523. [CrossRef]
31. Suarez-Bonnet, E.; Carvajal, M.; Mendez-Ramirez, I.; Castillo-Urueta, P.; Cortés-Eslava, J.; Gómez-Arroyo, S.; Melero-Vara, J.M. Aflatoxin (B_1, B_2, G_1, and G_2) contamination in rice of Mexico and Spain, from local sources or imported. *J. Food Sci.* **2013**, *78*, T1822–T1829. [CrossRef]
32. Chan, D.; Macdonald, S.J.; Boughtflower, V.; Brereton, P. Simultaneous determination of aflatoxins and ochratoxin A in food using a fully automated immunoaffinity column clean-up and liquid chromatography–fluorescence detection. *J. Chromatogr. A* **2004**, *1059*, 13–16. [CrossRef]
33. Gong, Y.; Yang, T.; Xiaosong, M.O.; Mo, X.; Zhang, H.; Huang, W.; Lu, T.; Zhang, D.; Wang, H. Preparation of Immunoaffinity Column for AFB_1 with Monoclonal Antibodies Immobilized on Protein A-Sepharose. *Food Sci.* **2015**, *21*, 193–198.
34. Göbel, R.; Lusky, K. Simultaneous determination of aflatoxins ochratoxin A, and zearalenone in grains by new immunoaffinity column/liquid chromatography. *J. AOAC Int.* **2004**, *87*, 411–416. [CrossRef] [PubMed]
35. Krska, R. Performance of modern sample preparation techniques in the analysis of Fusarium mycotoxins in cereals. *J. Chromatogr. A* **1998**, *815*, 49–57. [CrossRef]
36. Li, P.; Zhang, Q.; He, T.; Zhang, Z.; Ding, X. Immunosorbent and Immunoaffinity Column for Aflatoxin M1 Nanobody and Preparation Method Thereof. U.S. Patent 2015/0276729 A1, 6 September 2016.
37. Xie, J.; Sun, Y.; Zheng, Y.; Wang, C.; Sun, S.; Li, J.; Ding, S.; Xia, X.; Jiang, H. Preparation and application of immunoaffinity column coupled with dcELISA detection for aflatoxins in eight grain foods. *Food Control* **2017**, *73*, 445–451. [CrossRef]
38. Ediage, E.N.; Van Poucke, C.; De Saeger, S. A multi-analyte LC–MS/MS method for the analysis of 23 mycotoxins in different sorghum varieties: The forgotten sample matrix. *Food Chem.* **2015**, *177*, 397–404. [CrossRef]
39. Ardic, M.; Karakaya, Y.; Atasever, M.; Durmaz, H. Determination of aflatoxin B_1 levels in deep-red ground pepper (isot) using immunoaffinity column combined with ELISA. *Food Chem. Toxicol.* **2008**, *46*, 1596–1599. [CrossRef]
40. Arroyo-Manzanares, N.; Gámiz-Gracia, L.; García-Campaña, A.M. Determination of ochratoxin A in wines by capillary liquid chromatography with laser induced fluorescence detection using dispersive liquid–liquid microextraction. *Food Chem.* **2012**, *135*, 368–372. [CrossRef]
41. Goryacheva, I.Y.; Karagusheva, M.A.; Van Peteghem, C.; Sibanda, L.; Saeger, S.D. Immunoaffinity pre-concentration combined with on-column visual detection as a tool for rapid aflatoxin M1 screening in milk. *Food Control* **2009**, *20*, 802–806. [CrossRef]
42. Juan, C.; Zinedine, A.; Molto, J.C.; Idrissi, L.; Manes, J. Aflatoxins levels in dried fruits and nuts from Rabat-Salé area, Morocco. *Food Control* **2008**, *19*, 849–853. [CrossRef]
43. Goryacheva, I.Y.; Saeger, S.D.; Delmulle, B.; Lobeau, M.; Eremin, S.A.; Barna-VetrÓ, I.; Van Peteghem, C. Simultaneous non-instrumental detection of aflatoxin B_1 and ochratoxin A using a clean-up tandem immunoassay column. *Anal. Chim. Acta* **2007**, *590*, 118–124. [CrossRef]
44. Yang, T.; Lv, Y.; Zhang, D.; Tang, Y.; Yuan, Y.; Zhang, H.; Wang, H. Preparation and Application of Immunoaffinity Column for Multi-mycotoxins. *Chin. J. Anal. Chem.* **2016**, *44*, 1243–1249.
45. Hsie, A.W.; Brimer, P.A.; Mitchell, T.J.; Gosslee, D.G. The dose-response relationship for ethyl methanesulfonate-induced mutations at the hypoxanthine-guanine phosphoribosyl transferase locus in Chinese hamster ovary cells. *Somat. Cell Mol. Genet.* **1975**, *1*, 247–261. [CrossRef] [PubMed]
46. Tergel; Tai, D.; Liu, G.; Daorna; Wang, R.; Guo, H.; Li, Y.; Li, Y. Establishment of Hypoxanthine Guanine Phosphoribosyl Transferase Mutant from Human Liver Cancer Cell Line HepG2. *China Anim. Husb. Vet. Med.* **2011**, *38*, 92–95.
47. Kufer, P.; Lutterbüse, R.; Baeuerle, P.A. A Revival of Bispecific Antibodies. *Trends Biotechnol.* **2004**, *22*, 238–244. [CrossRef] [PubMed]
48. Beatty, J.D.; Beatty, B.G.; Vlahos, W.G. Measurement of monoclonal antibody affinity by non-competitive enzyme immunoassay. *J. Immunol. Methods* **1987**, *100*, 173–179. [CrossRef]
49. Delmulle, B.; Saeger, S.D.; Adams, A.; Kimpe, N.D.; Van Peteghem, C. Development of a liquid chromatography/tandem mass spectrometry method for the simultaneous determination of 16 mycotoxins on cellulose filters and in fungal cultures. *Rapid Commun. Mass Spectrom.* **2006**, *20*, 771–776. [CrossRef] [PubMed]
50. Junsai, T.; Poapolathep, S.; Sutjarit, S.; Giorgi, M.; Zhang, Z.; Logrieco, A.F.; Li, P.; Poapolathep, A. Determination of Multiple Mycotoxins and Their Natural Occurrence in Edible Vegetable Oils Using Liquid Chromatography-Tandem Mass Spectrometry. *Foods* **2021**, *10*, 2795. [CrossRef] [PubMed]
51. Akinyemi, M.O.; Braun, D.; Windisch, P.; Warth, B.; Ezekiel, C.N. Assessment of multiple mycotoxins in raw milk of three different animal species in Nigeria. *Food Control* **2022**, *131*, 108258. [CrossRef]
52. Zhang, W.; Liu, Y.; Liang, B.; Zhang, Y.; Zhong, X.; Luo, X.; Huang, J.; Wang, Y.; Cheng, W.; Chen, K. Probabilistic risk assessment of dietary exposure to aflatoxin B_1 in Guangzhou, China. *Sci. Rep.* **2020**, *10*, 7973. [CrossRef]

53. Fan, K.; Ji, F.; Xu, J.; Qian, M.; Duan, J.; Nie, D.; Tang, Z.; Zhao, Z.; Shi, J.; Han, Z. Natural Occurrence and Characteristic Analysis of 40 Mycotoxins in Agro-Products from Yangtze Delta Region. *Sci. Agric. Sin.* **2021**, *54*, 2870–2884.

foods

MDPI

Article

Hapten Synthesis and Monoclonal Antibody Preparation for Simultaneous Detection of Albendazole and Its Metabolites in Animal-Origin Food

Shibei Shao, Xuping Zhou, Leina Dou, Yuchen Bai, Jiafei Mi, Wenbo Yu, Suxia Zhang, Zhanhui Wang and Kai Wen *

Beijing Laboratory for Food Quality and Safety, Beijing Key Laboratory of Detection Technology for Animal-Derived Food Safety, College of Veterinary Medicine, China Agricultural University, Beijing 100193, China; shaoshibeil@cau.edu.cn (S.S.); 20098297@bua.edu.cn (X.Z.); b20193050410@cau.edu.cn (L.D.); BS20193050465@cau.edu.cn (Y.B.); mijiafei@cau.edu.cn (J.M.); yuwenbo@cau.edu.cn (W.Y.); suxia@cau.edu.cn (S.Z.); wangzhanhui@cau.edu.cn (Z.W.)

* Correspondence: wenkai@cau.edu.cn

Abstract: Albendazole (ABZ) is one of the benzimidazole anthelmintics, and the overuse of ABZ in breeding industry can lead to drug resistance and a variety of toxic effects in humans. Since the residue markers of ABZ are the sum of ABZ and three metabolites (collectively referred to as ABZs), albendazole-sulfone ($ABZSO_2$), albendazole-sulfoxide (ABZSO), and albendazole-2-amino-sulfone ($ABZNH_2SO_2$), an antibody able to simultaneously recognize ABZs with high affinity is in urgent need to develop immunoassay for screening purpose. In this work, an unreported hapten, 5-(propylthio)-1H-benzo[d]imidazol-2-amine, was designed and synthesized, which maximally exposed the characteristic sulfanyl group of ABZ to the animal immune system to induce expected antibody. One monoclonal antibody (Mab) that can simultaneously detect ABZs was obtained with IC_{50} values of 0.20, 0.26, 0.77, and 10.5 μg/L for ABZ, $ABZSO_2$, ABZSO, and $ABZNH_2SO_2$ in ic-ELISA under optimized conditions respectively, which has been never achieved in previous reports. For insight into the recognition profiles of the Mab, we used computational chemistry method to parameterize cross-reactive molecules in aspects of conformation, electrostatic fields, and hydrophobicity, revealing that the hydrophobicity and conformation of characteristic group of molecules might be the key factors that together influence antibody recognition with analytes. Furthermore, the practicability of the developed ic-ELISA was verified by detecting ABZs in spiked milk, beef, and liver samples with recoveries of 60% to 108.8% and coefficient of variation (CV) of 1.0% to 15.9%.

Keywords: albendazole; metabolites; hapten design; antibody; immunoassay; computational chemistry

Citation: Shao, S.; Zhou, X.; Dou, L.; Bai, Y.; Mi, J.; Yu, W.; Zhang, S.; Wang, Z.; Wen, K. Hapten Synthesis and Monoclonal Antibody Preparation for Simultaneous Detection of Albendazole and Its Metabolites in Animal-Origin Food. *Foods* **2021**, *10*, 3106. https://doi.org/10.3390/foods10123106

Academic Editor: Paolo Polidori

Received: 9 November 2021
Accepted: 1 December 2021
Published: 14 December 2021

Publisher's Note: MDPI stays neutral with regard to jurisdictional claims in published maps and institutional affiliations.

1. Introduction

Albendazole (ABZ, shown in Figure 1a), one of benzimidazoles, is an effective anthelmintic and often used to control soil-transmitted helminth infection in humans and animals. ABZ is usually the first choice for treatment of parasitic diseases, such as cystic echinococcosis and alveolar echinococcosis [1–3], and for eliminating lymphatic filariasis in endemic areas [4–6]. After administration of ABZ, the kinds of residues that can be monitored depend on the route of administration, target tissue or the detection time after treatment. At early periods, the most likely residues are albendazole sulfoxide (ABZSO) and albendazole sulfone ($ABZSO_2$), while albendazole 2-amino sulfone ($ABZNH_2SO_2$) may be the most persistent residue in tissue at longer withdrawal periods [7,8]. Residues of ABZ in animal food can lead to embryonic toxicity for consumer, as well as teratogenic and mutagenic effects [9,10]. The European Union (EU) has set the maximum residue limits (MRLs) for ABZs (sum of ABZ, ABZSO, $ABZSO_2$, and $ABZNH_2SO_2$, Figure 1a) at

100 μg/kg (L) in milk/muscle/fat, 500 μg/kg in kidney, and 1000 μg/kg in liver [11]. The U.S. FDA has set different residue limits in different samples, for example, 200 μg/kg in liver, 50 μg/kg in muscle [12]. The Ministry of Agriculture and Rural Affairs of China has set the MRLs (GB 31650-2019) for ABZs (sum of ABZ, ABZSO, $ABZSO_2$, and $ABZNH_2SO_2$) at 100 μg/L in milk, and for $ABZNH_2SO_2$ at 100–5000 μg/kg in edible tissues of food-producing species [13]. To ensure food safety, various instrumental methods have been established for determining ABZs in animal-derived food [14–19]. Although these methods are highly sensitive, they rely heavily on expensive instruments and skilled personnel, which often cannot meet the current urgent need for fast screening. Thus, there is great interest in the development of accurate and fast methods.

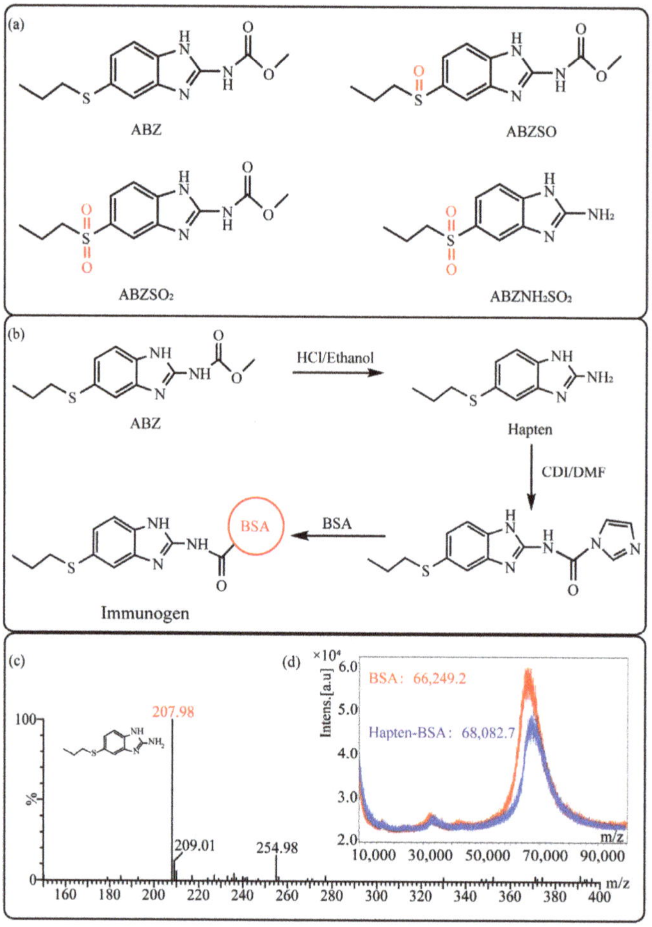

Figure 1. Chemical structures and synthesis route used in the study: (**a**) Chemical structure of four ABZs. (**b**) Synthesis route of hapten and immunogen. (**c**) Mass spectrometry characterization of hapten and (**d**) Matrix-assisted laser desorption/ionization time-of-flight mass spectrometry characterization of hapten- bovine serum albumin (BSA) conjugates.

Immunoassays, based on antibody-antigen recognition, offer a convenient and rapid alternative to instrumental methods and are currently the most widely used screening methods in food safety. Though immunoassay has many advantages including high-throughput, rapidity, and on-line detection, few studies have been conducted on immunoassay development due to the unavailability of antibody with high affinity and specificity to all ABZs, especially $ABZNH_2SO_2$. There have been only several reports of antibodies production for benzimidazoles including some of ABZs (Table 1), but all of the antibodies were of limited affinity to $ABZNH_2SO_2$, and the non-specificity binding to various benzimidazoles greatly constrained its application in detection of ABZs. Hapten design is the key to producing antibody with desired affinity and specificity. In previous studies, the strategy of maintaining the common moiety of benzimidazoles, carbamate group as the epitope [20–23] was used to obtain a broad-spectrum antibody. To the best of our knowledge, no report describing the production of antibody able to sensitively and specifically recognize all four ABZs has been reported despite of the great demand. In the present study, we described the synthesis of one new hapten, production of a Mab with highly specificity, exploration of antibody recognition mechanism, and development of an indirect competitive ELISA (ic-ELISA) for the determination of ABZ and its metabolites in milk and tissue simultaneously with improved sensitivity.

Table 1. Specificities of reported antibodies against ABZs in the literatures.

Compounds	Haptens&IC_{50} (μg/L)			
	[22]	[21]	[20]	This study
	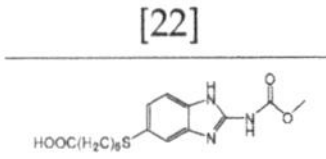	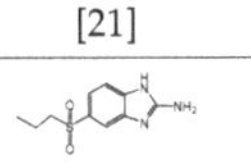	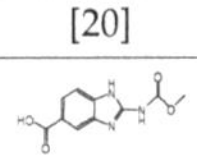	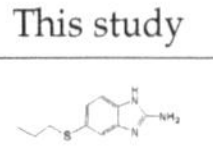
ABZ	1.4	>10,000	0.66	0.2
$ABZSO_2$	1.8	1253.2	5.34	0.26
ABZSO	1.5	2241.4	2.91	0.77
$ABZNH_2SO_2$	>10,000	85.2	>1000	10.5
Carbendazim	- [1]	-	14.84	>312.5
Fenbendazole	3.8	>10,000	0.75	1.68
Fenbendazole sulfone	8.3	-	6.27	-
Flubendazole	0.63	>10,000	0.37	3.68
Mebendazole	2.4	>10,000	0.3	4.14
Oxfendazole	0.62	>10,000	19.99	>312.5
Oxibendazole	1.4	>10,000	0.64	2.29
Parbendazole	-	-	1.13	-
Cambendazole	>100	-	>1000	-
Thiabendazole	>100	>10,000	-	>312.5

[1] Not mentiond or not detected.

2. Materials and Methods

2.1. Reagents and Materials

ABZ, ABZSO, $ABZSO_2$, $ABZNH_2SO_2$, fenbendazole, flubendazole, mebendazole, oxibendazole, oxfendazole, triclabendazole, carbendazim, and thiabendazole were purchased from J&K Chemical Technology (Beijing, China). Carbonyldiimidazole and other chemical reagents were supplied by Sinopharm Chemical Reagent (Beijing, China). Ovalbumin (OVA), bovine serum albumin (BSA), hypoxanthine aminopterin thymidine (HAT), complete and incomplete Freund's adjuvant, poly (ethylene glycol) (PEG) 1500, and fetal bovine serum were acquired from Sigma–Aldrich (St. Louis, MO, USA). Goat anti-mouse IgG (HRP labeled) was purchased from Jackson Immuno Research (West Grove, PA, USA). Cell culture medium (DMEM) was supplied by Thermo Fisher Scientific (Waltham, MA, USA). TMB (3,3′,5,5′-tetramethyl benzidine) substrate solution and hydrogen peroxide (H_2O_2) were purchased from Beyotime (Shanghai, China). Distilled water used in this study was obtained from a Milli-Q purification system (Bedford, MA, USA). Microplates for ELISA were acquired from Costar (Cambridge, MA, USA). Flat-bottomed high-binding polystyrene cell culture plates were obtained from Corning Life Sciences (New York, NY, USA). Balb/c mice were supplied by Beijing Vital River Laboratory Animal Technology

(Beijing, China). All of the buffers used in the immunoassay have been listed in Supplementary Materials.

2.2. Preparation of hapten and conjugates

2.2.1. Synthesis and Identification of Hapten

The hapten synthesis route is shown in Figure 1b and briefly described as follows: 500 mg of ABZ was firstly dissolved in 5.0 mL of ethanol in a round bottom flask with a magnetic stirrer, and then 10 mL of hydrochloric acid was added and heated to 80 °C for 30 min with stirring and monitored by thin-layer chromatography. After reaction, the pH value of the mixture was adjusted to 9.0 by using 2 M of sodium hydroxide solution, and extracted twice with 30 mL of ethyl acetate. The organic phase was combined and dried by using 2.0 g water-free sodium sulfate and the precipitate were removed. Then, 1.5 g of 100–200 mesh silicon was added to the organic phase and then dried for subsequent column chromatography (40 g 200–300 mesh chromatography silicone, eluted by petroleum ether: ethyl acetate 4:1). Finally, the hapten was obtained and vacuum-dried for mass spectrometry confirmation as shown in Figure 1c.

2.2.2. Synthesis and Identification of Immunogen and Coating Antigen

Amino groups of the acquired hapten were conjugated to carrier proteins by using carbonyldiimidazole as coupling reagent. Firstly, 9.2 mg of hapten was dissolved in 1.5 mL of DMF and stirred at 200 rpm for 10 min, then 7.6 mg carbonyldiimidazole was added, and the solution was stirred at room temperature (500 rpm) for 3 h to obtain activated hapten for subsequent coupling with carrier protein. Next, 50 mg of BSA was dissolved in 3.5 mL of 0.1 M of sodium bicarbonate solution, stirred at 200 rpm for 10 min, then cooled down via ice-bath. The previous activated hapten was then dropped into the protein solution at a rate of 1.0 mL/min under 1000 rpm stirring, then mixed at 500 rpm for 24 h. The reaction products were dialyzed for 3 days against phosphate buffer solution (PBS, 0.01 M, pH 7.2) under 4 °C. Finally, the product was centrifuged at 5000 rpm for 6 min to harvest the purified supernatant and stored at −20 °C until use. Immunogen was characterized by matrix-assisted laser desorption/ionization time-of-flight mass specTrometry (MALDI-TOF-MS, see Figure 1c) and the conjugation ratio was calculated as follows:

$$\text{Conjugation ratio} = (\text{M (conjugates)} - \text{M (BSA)})/\text{M (haptens)} \quad (1)$$

Coating antigen was synthesized using the above-mentioned procedure, except OVA was substituted for BSA.

2.3. Production of Monoclonal Antibody

All animal experiments were conducted in strict accordance with Chinese laws and guidelines approved by the animal ethics committee of China agricultural university. Eight Balb/c female mice aged 8 weeks were immunized with immunogen (diluted in PBS to 1.0 mg/mL) at a dose of 100 μg per mouse on an identical schedule. For primary immunization, mice were injected subcutaneously with a fully emulsified mixture of equal volumes complete Freund's adjuvant and prepared immunogen. For enhancement, mice were immunized with an emulsified mixture of immunogen and incomplete Freund's adjuvant every 3 weeks. A total of 4 immunizations were administered with the last one given via intraperitoneal injection without the adjuvant. To better monitor the serum titer and specificity by ic-ELISA, serum was collected 7–14 days after each immunization, according to results of a dynamics study of the antibody-mediated immune response [24]. Four days after the last boost immunization, mice spleen cells were separated and fused with PEG 1500 pre-treated sp2/0 myeloma cells to prepare hybridomas according to procedures described previously [25–27]. The fused cells were cultured in HAT medium for 7 days and were screened by testing the supernatant using ic-ELISA to determine the binding ability. The positive and highly sensitive hybridomas were obtained after subcloning four times using the limiting dilution method. Finally, the hybridomas were intraperitoneally injected into mice, and the ascites collected from mice were extracted and purified with saturated ammonium sulfate to obtain purified Mabs.

2.4. *Development and Optimization of ic-ELISA*

An ic-ELISA was established under the following optimized assay conditions: microplates were coated with 100 μL of coating antigen dissolved in 0.05 M carbonate buffer and incubated at 37 °C for 2 h. The plates were then washed three times for subsequent blocking. A volume of 150 μL/well of blocking buffer was added and incubated at 37 °C for 1.5 h, after which the buffer was removed. For the competitive step, both 50 μL of competitor and 50 μL of antibody working solution were pipetted into each well and incubated at 37 °C for 30 min. The plates were then washed three times, 100 μL/well of goat-anti-mouse IgG-HRP diluted in PBS (1:5000) was added, and the plates were incubated at 37 °C for 30 min. The plates were washed as above, 100 μL/well of newly prepared substrate solution was added, and the plates were incubated at 37 °C for 15 min. Finally, the chromogenic reaction was terminated with 50 μL/well H_2SO_4 (2 M). Optical density (OD) values at 450 nm were measured with a Multiskan FC machine by Thermo Scientific (Shanghai, China). The OD450 values were plotted against the analyte concentration on a logarithmic scale, and the generated sigmoidal curve was mathematically fitted to a four-parameter logistic equation using the OriginPro 8.5 software (OriginLab Corporation, Northampton, MA, USA).

$$Y = (A - D)/(1 + (X/C)^B) + D \tag{2}$$

where A = response at high asymptote, B = the slope factor, C = concentration corresponding to 50% specific binding (IC_{50}), D = response at low asymptote, and X = the calibration concentration.

Several physicochemical factors were optimized to improve the performance of ic-ELISA, including pH and ionic strength of working solution.

2.4.1. Effect of pH

The effect of varying the pH on the ic-ELISA was tested by dissolving the analytes and Mabs in PBS buffer at a specified pH and adding them to the antigen coated plates in equal volumes (50 μL/well). The pH values of 5.5, 6.5, 7.0, 7.4, and 8.5 were tested in the ic-ELISA incubation step with all other parameters of the assay fixed.

2.4.2. Effect of Ionic Strength

The effect of ionic strength of assay buffer on the ic-ELISA performance was studied using different NaCl concentrations of 0.05, 0.1, 0.15, 02, 0.4, and 0.8 M in 0.01 M of phosphate buffer, respectively. The effects of these salt concentrations were evaluated on the ic-ELISA by comparing the ABZ competition curves measured with each buffer at pH 7.4.

2.5. *Cross-Reactivities and Computational Chemistry Analysis*

The specificity of the developed ELISA was assessed using cross-reactivity (CR) determined under optimal conditions. The CRs of ABZs and other widely used benzimidazole anthelmintics, such as fenbendazole, oxibendazole, mebendazole, flubendazole, oxfendazole, triclabendazole, carbendazim, and thiabendazole, were calculated according to the following equation:

$$CR = IC_{50}\ (ABZ, \mu g/L)/IC_{50}\ (analytes, \mu g/L) \times 100\% \tag{3}$$

The computational chemistry method, which can provide electrostatic potential and conformational information of the molecule regarding antibody recognition, was employed here to study the CRs of immunoassays and the binding interactions between small molecules and antibody. All 3D structures were built in Gaussian 09 (Gaussian Inc., Wallingford, CT, USA) and then optimized by the density functional theory calculations with the M06-2X density functional and TZVP basis set. Basing on the lowest energy conformations, molecular alignments were materialized by molecular overlay modules in Discovery Studio 2019 (Accelrys Software, Inc., San Diego, CA, USA). The degree of molecular superposition was measured by alignment root-mean-square-deviation (RMSD). The quantitative molecular surface analysis in the Multiwfn software package was then applied together with VMD to analyze molecular electrostatic potential (ESP) and map ESP on van der Waals surface [28,29]. The molecule volume, and total polar surface area (TPSA) were extracted using the Multiwfn 3.7(dev) code. Dipole moment (μ) was extracted from the Gaussian output file. The Log P was obtained using ChemDraw (PerkinElmer, Waltham, MA, USA). By analyzing steric and electrostatic contour maps of the region around the molecule with respect to changes in affinity, the structure–activity relationship between the drugs and the Mab was studied.

2.6. Sample Preparation

Negative samples (milk, muscle, and liver of bovine) were obtained from the National Reference Laboratory for Veterinary Drug Residues (Beijing, China). Tissue samples (muscle and liver) were spiked with $ABZNH_2SO_2$, and milk was spiked with ABZ/ABZSO/$ABZSO_2$/$ABZNH_2SO_2$. Three grams (milliliters) homogenized sample were weighed into a 50-mL centrifuge tube with 1.0 g of sodium sulfate for pretreatment. Two milliliters of 50% NaOH (only tissue sample) and 9.0 mL of ethyl acetate were then added to the above tube. The mixture was vortexed for 5 min and centrifuged at 4000 rpm for 10 min. Next, 4.5 mL of the supernatant was transferred into a new tube and dried at 60 °C under a nitrogen stream. One milliliter of n-hexane and 0.5 mL of acetonitrile were added to the dried tube for cleaning and resuspension and followed by centrifugation at 4000 rpm. The upper n-hexane and commixture was discarded then, and 0.1 mL of the remaining liquid was pipetted into a new centrifuge tube and mixed with 0.9 mL PBS (0.01 M) (for milk samples) or 1.9 mL PBS (for tissue samples) for detection. The limit of detection (LOD) was determined as the 10% inhibition concentration (IC_{10}) calculated from calibration curves. All measurements were completed in triplicate.

3. Results and Discussion

3.1. Hapten Design and Characterization

Hapten structure determines the specificity and affinity of the resulting antibody [30]. To produce the antibody that could recognize ABZs, the characteristic structure of these compounds should be maximumly exposed to immune system. Previous studies have designed hapten by maintaining the carbamate group, which is a common structure in major benzimidazoles, to prepare antibodies with broad-spectrum specificities [20]. Some of the obtained antibodies could detect ABZ with affinity of 0.66 to 1.4 μg/L (Table 1), but showed poor affinity to ABZ metabolites especially $ABZNH_2SO_2$, which are critical residue markers of ABZ. Even worse, the specificities of these antibodies were insufficient because of the recognition of many other benzimidazoles with CRs of 3% to 88%. A similar result was also observed in another study, in which the $ABZNH_2SO_2$ was directly used as hapten to produce antibodies against $ABZNH_2SO_2$ [21].

In this study, we designed a new hapten of ABZs as shown in Figure 1b and Table 1. The feature structure of ABZ, the sulfanyl group, was maximumly exposed to immune system to induce an antibody with specific recognition of ABZs by conjugating with proteins at the carbamate moiety. Besides the increasing in complexity of epitope (often means stronger antigenicity) relative to previous ones, another key point in this design is that the retained sulfanyl group of ABZ could better maintain the characteristic structures of ABZs, which are structurally different from other benzimidazoles. The hapten was identified by High performance liquid chromatography-tandem mass spectrometry (HPLC-MS/MS, as shown in Figure 1c) (Agilent Technologies, Santa Clara, CA, USA) and ^1H nuclearmagnetic resonance spectrometry (NMR, see Figure S1 in Supplementary Materials) (Bruker, Rheinstetten, Germany), and the molecular ions (m/z) of hapten were 207.98, indicating that the hapten was successfully obtained. The hapten-BSA conjugate was then used as the immunogen, while hapten-OVA as the coating antigen. The immunogen was characterized by MALDI-TOF-MS (Bruker, Rheinstetten, Germany). The correct molecular weight of the hapten molecular ion and conjugate were observed (see Figure 1d), demonstrating that the hapten had been conjugated to the carrier protein, and the calculated conjugation ratio of hapten to BSA was 8.8:1.

3.2. Development and Optimization of the ic-ELISA

Four ABZs, including ABZ, $ABZSO_2$, ABZSO, and $ABZNH_2SO_2$, were used to determine the affinity (expressed by IC_{50}) and specificity (expressed by CR) of antibodies using homologous coating antigen. The results indicated that the obtained antibodies 12F12, 12H3, and 4E1 could recognize ABZs with varied affinities (IC_{50} values of antisera were from 0.12 μg/L to beyond 100 μg/L). As shown in Figure 2a, the antibody 4E1 exhibited relatively higher affinity to ABZ and ABZSO, but relatively worse recognition towards $ABZNH_2SO_2$; 12H3 had high affinity to three ABZs but unfavorable affinity to $ABZNH_2SO_2$; while the antibody 12F12 showed high affinity to all four ABZs and was selected for further optimization.

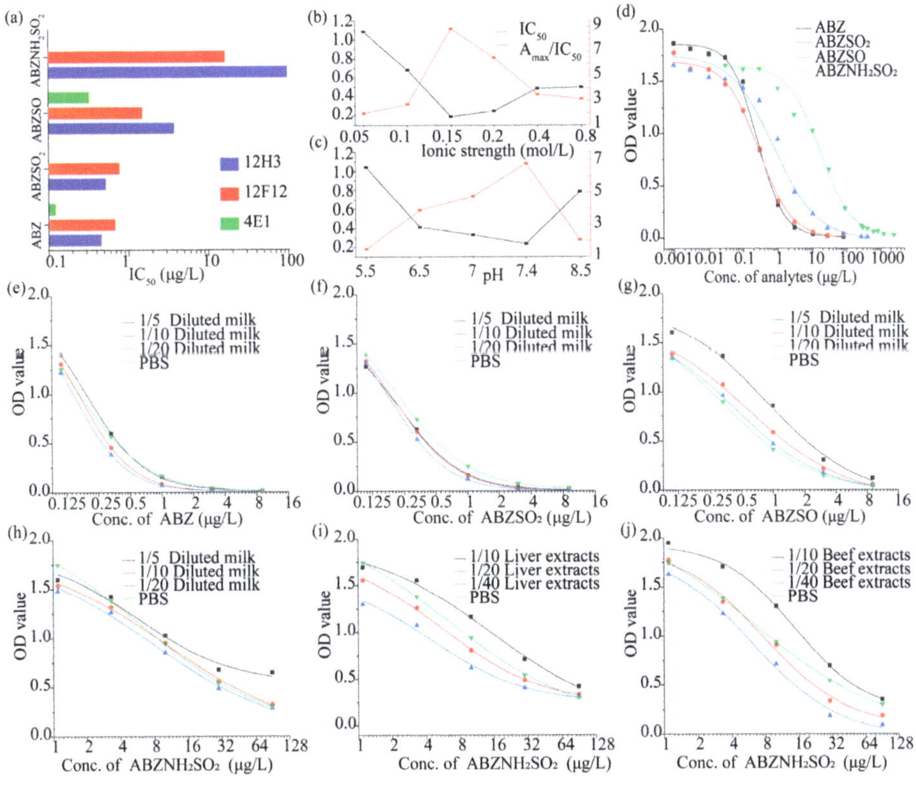

Figure 2. Optimization of ic-ELISA. The IC_{50} was calculated when the A_{max} (Optical density (OD) value of negative control) ranged from 1.0 to 2.0. (**a**) Screening of Mab. Each Mab was estimated using four ABZs separately. The effect of (**b**) pH value and (**c**) ionic strength in the ic-ELISA were evaluated using Mab 12F12 and ABZ. (**d**) Standard curves of ic-ELISA for ABZs. (**e–j**) the calibration curves of the ic-ELISA for (**e**) ABZ, (**f**) $ABZSO_2$, (**g**) ABZSO, (**h**) $ABZNH_2SO_2$ in phosphate buffer solution (PBS) and milk, and of $ABZNH_2SO_2$ in extracted (**i**) bovine liver and (**j**) beef. Each value represents the average of three independent replicates.

The development and optimization procedure were performed according to previous studies [25,26]. Ionic strength and pH value of working buffer were further optimized. The A_{max} (means the OD value of negative control)/IC_{50} ratio was introduced here to be a criterion for optimization, and a higher ratio value meant a higher sensitivity. It can be seen that the ratio of A_{max}/IC_{50} was highest at an ionic strength of 0.15 mol/L (Figure 2b) and pH 7.4 (Figure 2c). Thus, the optimum conditions (pH 7.4 and 0.15 mol/L NaCl in PB buffer, namely, 0.01 M PBS) were used in subsequent experiments.

The sensitivity of the ic-ELISA was characterized by IC_{50} values from standard curves under the optimized conditions, Standard curves of ABZ/$ABZSO_2$/ABZSO were established at 27.0, 9.0, 3.0, 1.0, 0.33, 0.11, 0.037, 0.012, 0.004, 0.001, and 0 μg/L in PBS, standard curves of $ABZNH_2SO_2$ were established at 270, 90, 30, 10, 3.33, 1.11, 0.37, 0.12, 0.04, 0.01, and 0 μg/L in PBS. As shown in Figure 2d, the developed ic-ELISA based on antibody 12F12 could detect ABZ, $ABZSO_2$, and ABZSO with IC_{50} values below 1.0 μg/L (0.20, 0.26, and 0.77 μg/L), and $ABZNH_2SO_2$ with IC_{50} values of 10.5 μg/L in PBS buffer, which is of the highest affinity so far. The results shown that the sensitivities of the developed ic-ELISA were times better than those of other reported immunoassays for ABZs (Table 1).

3.3. *Cross-Reactivities and Structure-Activity Relationship Study by Computational Chemistry*

In the process of antibody-antigen recognition, cross-reactivity would arise if the configuration of the antigen matches the active pocket of the antibody [31,32]. As shown in Table 2, the ic-ELISA exhibited varied CRs with ABZ (100%), $ABZSO_2$ (76.9%), ABZSO (26.0%), $ABZNH_2SO_2$ (1.9%), and other often-used benzimidazoles (fenbendazole, flubendazole, mebendazole, oxibendazole) less than 8%. The relative lower CR value of $ABZNH_2SO_2$ (CR = 1.9%) can easily draw our attention, which shares similar structure with $ABZSO_2$ (CR = 76.9%), except for the carbamate group on the right of molecule in the case of $ABZSO_2$. As the sulfanyl group of hapten was exposed to immune system as a characteristic epitope in primary hapten design, it would be the site for the antibody recognition. While the remote amide group, which formed after the hapten was coupled to carrier protein, should be masked by carrier protein and contribute much less for antibody recognition. However, the recognition ability of the obtained Mab seemed not in accordance with expectation by analyzing only 2D dimension due to the relative poorer recognition of $ABZNH_2SO_2$ without a methoxyamide or carbamate group. Nevertheless, the benzimidazoles having complex (from a 2D view) side chain groups or phenyl groups (fenbendazole, flubendazole, mebendazole, oxibendazole, Table 2) surprisingly showed higher CRs than $ABZNH_2SO_2$. Though the ic-ELISA was with high sensitivity for detection of all ABZs and far better than previous study, it was appealing to further study on the CR data, and there was considerable interest in understanding at 3D level as well as quantitative analysis. Therefore, computational chemistry analysis was used to provide 3D conformations and quantitative information, such as configuration, electrostatic potential, and hydrophobicity of molecules, to investigate the binding of antibody to compounds. Based on their optimized lowest energy conformations of ABZs and other benzimidazoles that showed cross-reactivities, all molecules were aligned while ABZ was selected as the template. All hydrogen atoms were hidden in the optimized geometric structures for a concise view.

The hapten design of this work was based on the conjecture that the sulfanyl/sulfonyl moiety is the characteristic moiety of ABZs, thus should be retained to prepare ABZs antibodies. Figure 3a shows that when this moiety of the structure is changed, such as S atom (ABZ, in green) becomes O atom (oxibendazole, in yellow), chain alkyl group (ABZ, in green) becomes phenyl group (fenbendazole, in light blue), or both (flubendazole in brick-red, or mebendazole in grey), etc., at least a 10-fold decrease in CRs was observed, indicating the criticality of the sulfanyl/sulfonyl group of the structure. It can be seen in Figure 3a that all molecules are well coincided with ABZ, and the RMSD values of the alignments are from 2.3×10^{-3} of $ABZNH_2SO_2$ to 1.3×10^{-2} of $ABZSO_2$ and fenbendazole (Table 3), demonstrating the extent of the difference of other molecules with ABZ, which agreed with CR data on the whole. It can also be seen in Figure 3a, the substitution of carbon or oxygen atom for sulfur atom in flubendazole (in red), mebendazole (in gray), or oxibendazole (in gray) causes changes in torsion angle of the left group towards the benzimidazole ring comparing with that of ABZ (in red), which might be partly responsible for the 10-fold decrease of Mab recognition. The ABZSO (in blue), $ABZSO_2$ (in purple), and $ABZNH_2SO_2$ (in orange) show almost the same angle, but the difference between CRs is up to dozens of times. There shall be other factors which influence the recognition of antibody besides the contribution of molecular conformation.

Table 2. Experimental IC_{50} values and cross-reactivities of ABZs in optimized ic-ELISA.

Analytes		IC_{50} (μg/L)	CR (%)
ABZ	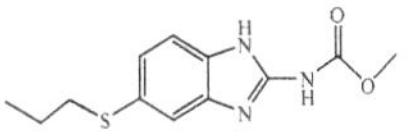	0.20	100
ABZSO2	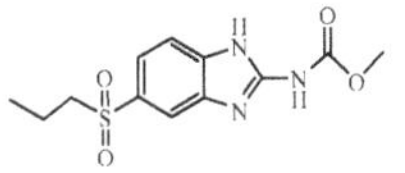	0.26	76.9
ABZSO	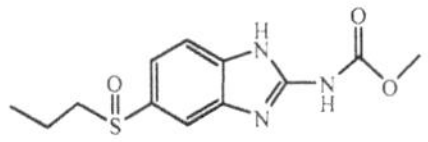	0.77	26.0
ABZNH2SO2	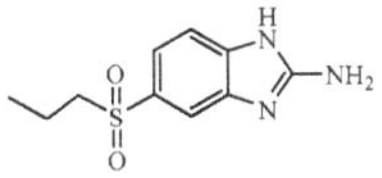	10.50	1.9
Fenbendazole	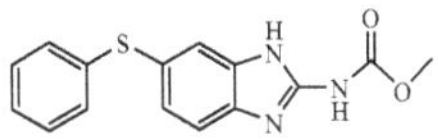	1.68	11.9
Flubendazole	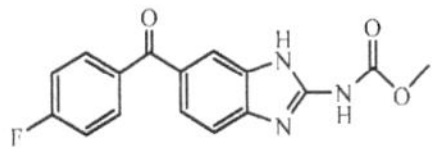	3.68	5.4
Mebendazole	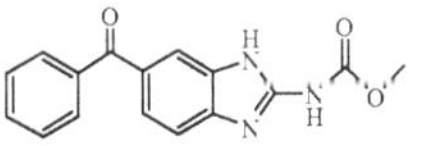	4.11	4.8
Oxibendazole	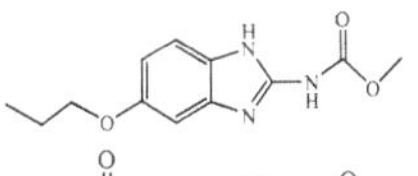	2.29	8.7
Oxfendazole	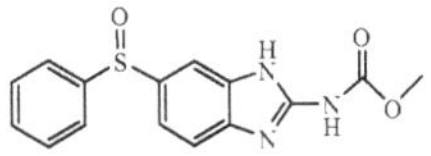	>312.5	<0.01
Triclabendazole	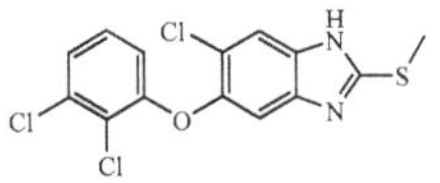	>312.5	<0.01
Carbendazim	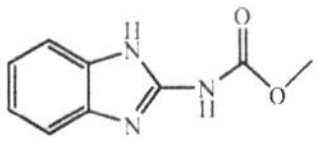	>312.5	<0.01
Thiabendazole	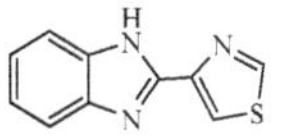	>312.5	<0.01

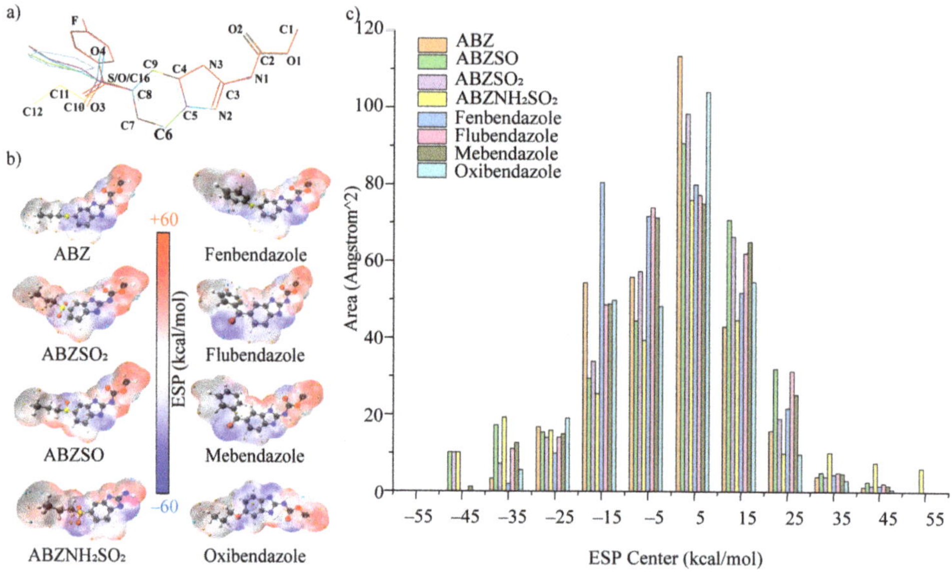

Figure 3. (**a**) The alignment of cross-reactive analytes with ABZ based on lowest energy conformations with all of the hydrogen atoms hidden. (**b**) Electrostatic potential energy of analytes. The negative potential areas are indicated in blue; red coloring indicates positive potential areas, and white indicates relatively neutral areas. The blue and golden globules on the surface represent the minima and maxima of ESP (kcal/mol) on the van der Waals surface. (**c**) The calculated surface area distribution in different ESP ranges on the van der Waals surface.

Table 3. Comparison of cross-reactivities and molecular descriptors of benzimidazoles.

Items	ABZ	$ABZSO_2$	ABZSO	$ABZNH_2SO_2$	Fenbendazole	Flubendazole	Mebendazole	Oxibendazole
CR (%)	100	76.9	26.0	1.9	11.9	5.4	4.8	8.7
MW	265.1	297.3	281.3	239.3	299.3	313.3	295.3	249.3
volume($Å^3$)	320.9	336.4	328.5	201.8	255.3	261.1	256.2	225.0
Log *P*	2.8	1.2	1.2	1.0	3.4	3.1	2.9	2.4
μ	4.1	7.6	4.7	6.5	4.8	6.6	6.8	1.8
TPSA ($Å^2$)	67.0	101.2	84.1	88.9	67.0	84.1	84.1	76.2
alignment RMSD	0	1.3×10^{-2}	1.1×10^{-2}	2.3×10^{-3}	1.3×10^{-2}	8.5×10^{-3}	8.4×10^{-3}	5.4×10^{-3}

ESP describes the potential energy of a proton placed at a point near the molecule. As different substituents may affect the electrostatic field, ESP analysis helps to visualize spatial regions of molecules and to analyze electrostatic interaction between antibody and antigen [30,33]. The ESP calculations displayed on van der Waals surfaces of global lowest energy conformation are shown in Figure 3b, positive potential energy is represented by red areas around the molecules, of which the maxima were marked as golden globule, and these areas are repulsive to a proton, negative potential energy is represented by blue areas on the molecules with the minima marked as light blue globule, and these areas are attractive to a proton. It can be seen that the presence of the oxygen atom causes more negative potential surface. With consideration of almost identical structure of $ABZNH_2SO_2$ (CR = 1.9%) and $ABZSO_2$ (CR = 76.9%), except for the methoxyamide group on the right of molecule in the case of $ABZSO_2$ and amino group in the case of $ABZSO_2NH_2$, less positive potential surface can be noticed in $ABZNH_2SO_2$, and the reduction of electronic effect associated with the lack of group might influence the recognition by Mab. It is clearly seen from Figure 3c that the ESP surface area distribution of molecules in different ESP ranges are similar on the whole but obviously varied locally. Cross-reactive analytes seem to have stronger electronic effect (positive or negative area) than ABZ, but do not generate better CRs. The results imply that the electrostatic potential energy might play a limited role in the antibody recognition or has been masked by other powerful factors.

Critical molecular descriptors were finally compared as shown in Table 3. Studies have shown that molecular parameters such as lipid-water partition coefficient (Log *P*, a widely used measure of hydrophobicity), dipole moments (μ, represents the polarity of molecule, the larger the polarity, the stronger the hydrophilicity), and topological polar surface area (TPSA) of analytes are important factors affecting recognition by antibodies [34–36], and hydrophobic force was believed to be the main driving force between small molecules and antibody [30,34]. In this study, Log *P* was found to be positively correlated with antibody affinity. The hydrophobicity decreases when the sulfur group in the molecule is replaced with a sulfone or a sulfoxide groups, resulting in a decrease in CR. Compared with ABZ, the hydrophobicity of $ABZNH_2SO_2$ greatly decreases, which is the lowest among all the analytes as well as the CR. Other compounds with strong hydrophobic groups do not show high CRs, which may be due to the co-working conformational factors. Some other physicochemical parameters that may influence the immunogenicity of haptens, including MW, SAs, and several hydrophobic parameters has also been calculated and summarized in Table 3, which shows no significant relationship with the recognition.

In general, hydrophobicity still dominates in the process of recognition in this study, and conformational factors play a partial role. Changes in the part of what we maintained as epitope in hapten design, the sulfanyl group, cause decrease in antibody recognition. This confirms the correctness of the hapten design strategy in this study, that is, the characteristic moiety of the target molecule should be maximally exposed.

3.4. *Matrix Effect and Recovery in Samples*

Milk, muscle, and liver samples from bovine were chosen to determine matrix effect and recovery. Direct dilution following sample extraction was used as a simple way to eliminate matrix effects for rapid screening purpose. The milk extracts were diluted 5/10/20 times with PBS and the tissue samples extracts were diluted 10/20/40 times with PBS. The standard curves of ABZs prepared in diluted samples were then compared with those in PBS to evaluate the matrix effects (Figure 2e–j).

Considering both A_{max} and IC_{50} performance, 10-fold dilution of milk samples and 20-fold dilution of tissue samples (beef and liver) in PBS were chosen to establish the calibration curves. The calculated LODs of ABZ, $ABZSO_2$, ABZSO, and $ABZNH_2SO_2$ were at 0.05, 0.05, 0.05, and 0.49 μg/L in milk samples, respectively, and the LODs of $ABZNH_2SO_2$ in beef and bovine liver samples were 1.12 and 0.56 μg/kg, which were below the MRL of ABZs. Negative foodstuffs were spiked with ABZs at concentrations of 0.5/2.0/10.0 or 10/30/90 μg/kg (L), and the average absorbance values were interpolated with calibration curves to determine recoveries. The linearity ranges (from IC_{20} to IC_{80}), recoveries, and CVs of ABZs in spiked milk samples are shown in Table 4, and that of $ABZNH_2SO_2$ in spiked tissue samples are shown in Table 5. As shown in Tables 4 and 5, the recoveries ranged from 60% to 108.8%, with CVs less than 15.9%, which confirmed that the ic-ELISA performed well in various matrices.

Table 4. Recoveries, CVs, LODs, and linearity range of ABZs in spiked milk samples using ic-ELISA. (n = 3).

	ABZs in Milk Samples				
	Spiked Level (μg/L)	Recovery (%)	CV (%)	LOD (μg/L)	Linearity Range (μg/L)
ABZ	0.5	87.5%	1.0%		
	2	82.1%	7.7%	0.05	0.08–0.4
	10	78.7%	6.9%		
$ABZSO_2$	0.5	94.2%	4.1%		
	2	71.4%	10.5%	0.05	0.08–0.5
	10	60.0%	6.8%		
ABZSO	0.5	105.1%	11.9%		
	2	98.7%	9.9%	0.05	0.1–2.5
	10	74.5%	7.9%		
$ABZNH_2SO_2$	10	108.0%	10.3%		
	30	76.6%	12.3%	0.50	1.5–73.5
	90	75.2%	2.9%		

Table 5. Recoveries, CVs, LODs, and linearity range of $ABZNH_2SO_2$ in spiked tissue samples using ic-ELISA. (n = 3).

	$ABZNH_2SO_2$ in Tissue Samples				
	Spiked Level (μg/kg)	Recovery (%)	CV (%)	LOD (μg/kg)	Linearity Range (μg/kg)
Beef	10	84.1%	2.4%		
	30	91.6%	4.9%	1.12	2.3–25.9
	90	74.2%	6.6%		
Liver	10	87.8%	4.5%		
	30	106.1%	9.1%	0.56	1.3–23.7
	90	108.8%	15.9%		

4. Conclusions

In this study, a new hapten was designed by exposing the characteristic sulfanyl group of ABZ as an epitope. Furthermore, one Mab 12F12 for simultaneously detection of ABZs including $ABZNH_2SO_2$ was prepared for the first time and with the highest affinity to date. The established ic-ELISA based on the Mab 12F12 presented times lower IC_{50} values than previously reported and was suitable for the screening of ABZs in foodstuffs. The molecular recognition mechanism was briefly explained via computational chemistry analysis and indicated that the hydrophobicity of molecules and conformational factors might be the key factors that affect the binding between antibody and ABZs in this study, which might offer guides for antibody preparation to reduce the cost of trial-and-error in subsequent research.

Supplementary Materials: The following are available online at https://www.mdpi.com/article/10.3390/foods10123106/s1, Figure S1: 1 H NMR spectra of hapten.

Author Contributions: Conceptualization, S.S. and Z.W.; Methodology, S.S., L.D. and J.M.; software, S.S. and Y.B.; validation, W.Y.; formal analysis, W.Y.; investigation, X.Z.; resources, X.Z.; data curation, S.S.; writing—original draft preparation, S.S.; writing—review and editing, S.S., L.D., Y.B., J.M. and W.Y.; visualization, S.S. and X.Z.; supervision, Z.W., K.W. and S.Z.; project administration, Z.W.; funding acquisition, Z.W. and K.W. All authors have read and agreed to the published version of the manuscript.

Funding: This research was funded by the Ministry of Science and Technology (MOST) for the National Key R&D Program of China, grant number 2018YFC1603500, Sanming Project of Medicine in Shenzhen (SZSM201611068), and supported by China Agriculture Research System of MOF and MARA.

Institutional Review Board Statement: Not applicable.

Informed Consent Statement: Not applicable.

Data Availability Statement: Not applicable.

Conflicts of Interest: The authors declare no conflict of interest.

References

1. Taylor, M.J.; Hoerauf, A.; Bockarie, M. Lymphatic filariasis and onchocerciasis. *Lancet* **2010**, *376*, 1175–1185. [CrossRef]
2. Gobbi, F.; Buonfrate, D.; Tamarozzi, F.; Degani, M.; Angheben, A.; Bisoffi, Z. Efficacy of high-dose albendazole with ivermectin for treating imported loiasis, Italy. *Emerg. Infect. Dis.* **2019**, *25*, 1574–1576. [CrossRef]
3. Macfarlane, C.L.; Budhathoki, S.S.; Johnson, S.; Richardson, M.; Garner, P. Albendazole alone or in combination with microfilaricidal drugs for lymphatic filariasis. *Cochrane. Db. Syst. Rev.* **2019**, *1*, CD003753. [CrossRef]
4. Budge, P.J.; Herbert, C.; Andersen, B.J.; Weil, G.J. Adverse events following single dose treatment of lymphatic filariasis: Observations from a review of the literature. *PLoS Negl. Trop. Dis.* **2018**, *12*, e0006454. [CrossRef]
5. Chami, G.F.; Bundy, D.A.P. More medicines alone cannot ensure the treatment of neglected tropical diseases. *Lancet Infect. Dis.* **2019**, *19*, 330–336. [CrossRef]
6. Pullan, R.L.; Halliday, K.E.; Oswald, W.E.; Mcharo, C.; Beaumont, E.; Kepha, S.; Witek-McManus, S.; Gichuki, P.M.; Allen, E.; Drake, T.; et al. Effects, equity, and cost of school-based and community-wide treatment strategies for soil-transmitted helminths in Kenya: A cluster-randomised controlled trial. *Lancet* **2019**, *393*, 2039–2050. [CrossRef]
7. Danaher, M.; De Ruyck, H.; Crooks, S.R.; Dowling, G.; O'Keeffe, M. Review of methodology for the determination of benzimidazole residues in biological matrices. *J. Chromatogr. B Analyt. Technol. Biomed. Life Sci.* **2007**, *845*, 1–37. [CrossRef] [PubMed]
8. Schulz, J.D.; Neodo, A.; Coulibaly, J.T.; Keiser, J. Pharmacokinetics of albendazole, Albendazole sulfoxide, and albendazole sulfone determined from plasma, Blood, Dried-blood spots, And mitra samples of hookworm-infected adolescents. *Antimicrob. Agents Chemother.* **2019**, *63*, e02489-18. [CrossRef] [PubMed]
9. Thomsen, E.K.; Sanuku, N.; Baea, M.; Satofan, S.; Maki, E.; Lombore, B.; Schmidt, M.S.; Siba, P.M.; Weil, G.J.; Kazura, J.W.; et al. Efficacy, safety, and pharmacokinetics of coadministered diethylcarbamazine, albendazole, and ivermectin for treatment of bancroftian filariasis. *Clin. Infect. Dis.* **2016**, *62*, 334–341. [CrossRef]
10. Ricken, F.J.; Nell, J.; Grüner, B.; Schmidberger, J.; Kaltenbach, T.; Kratzer, W.; Hillenbrand, A.; Henne-Bruns, D.; Deplazes, P.; Moller, P.; et al. Albendazole increases the inflammatory response and the amount of Em2-positive small particles of Echinococcus multilocularis (spems) in human hepatic alveolar echinococcosis lesions. *PLoS Negl. Trop. Dis.* **2017**, *11*, e0005636. [CrossRef] [PubMed]
11. The European Union. *Commission Regulation (EU) No 37/2010 of 22 December 2009 on Pharmacologically Active Substances and Their Classification Regarding Maximum Residue Limits in Foodstuffs of Animal Origin*; Official Journal of the European Union: Brussels, Belgium, 2010.
12. The, U.S. Food and Drug Administration. 21CFR556.34. Specific Tolerances for Residues of Approved and Conditionally Approved New Animal Drugs. Sec. 556.34 Albendazole. In *The Code of Federal Regulations*; U.S. Food and Drug Administration: Silver Spring, MD, USA, 2021.
13. The Ministry of Agriculture and Rural Affairs of the People's Republic of China. *National Food Safety Standard-Maximum Residue Limits for Veterinary Drugs in Foods*; GB 31650-2019; Standards Press of China: Beijing, China, 2019.
14. Li, S.; Liang, Q.; Ahmed, S.A.H. Simultaneous determination of five benzimidazoles in agricultural foods by core-shell magnetic covalent organic framework nanoparticle-based solid-phase extraction coupled with high-performance liquid chromatography. *Food Anal. Methods* **2020**, *13*, 1111–1118. [CrossRef]
15. Permana, A.D.; Tekko, I.A.; McCarthy, H.O.; Donnelly, R.F. New HPLC-MS method for rapid and simultaneous quantification of doxycycline, diethylcarbamazine and albendazole metabolites in rat plasma and organs after concomitant oral administration. *J. Pharm. Biomed. Anal.* **2019**, *170*, 243–253. [CrossRef]
16. Bach, T.; Bae, S.; D'Cunha, R.; Winokur, P.; An, G. Development and validation of a simple, fast, and sensitive LC/MS/MS method for the quantification of oxfendazole in human plasma and its application to clinical pharmacokinetic study. *J. Pharm. Biomed. Anal.* **2019**, *171*, 111–117. [CrossRef] [PubMed]
17. Xu, N.; Dong, J.; Yang, Y.; Liu, Y.; Yang, Q.; Ai, X. Development of a liquid chromatography-tandem mass spectrometry method with modified QuEChERS extraction for the quantification of mebendazole and its metabolites, albendazole and its metabolites, and levamisole in edible tissues of aquatic animals. *Food Chem.* **2018**, *269*, 442–449. [CrossRef] [PubMed]
18. Tejada-Casado, C.; Moreno-Gonzalez, D.; Lara, F.J.; Garcia-Campana, A.M.; Del Olmo-Iruela, M. Determination of benzimidazoles in meat samples by capillary zone electrophoresis tandem mass spectrometry following dispersive liquid-liquid microextraction. *J. Chromatogr. A* **2017**, *1490*, 212–219. [CrossRef]
19. Xia, X.; Wang, Y.; Wang, X.; Li, Y.; Zhong, F.; Li, X.; Huang, Y.; Ding, S.; Shen, J. Validation of a method for simultaneous determination of nitroimidazoles, benzimidazoles and chloramphenicols in swine tissues by ultra-high-performance liquid chromatography-tandem mass spectrometry. *J. Chromatogr. A* **2013**, *1292*, 96–103. [CrossRef]

20. Guo, L.; Wu, X.; Liu, L.; Kuang, H.; Xu, C. Gold nanoparticle-based paper sensor for simultaneous detection of 11 benzimidazoles by one monoclonal antibody. *Small* **2018**, *14*, 1701782. [CrossRef] [PubMed]
21. Peng, D.; Jiang, N.; Wang, Y.; Chen, D.; Liu, Z.; Yuan, Z. Development and validation of an indirect competitive enzyme-linked immunosorbent assay for the detection of albendazole 2-aminosulfone residues in animal tissues. *Food Agric. Immunol.* **2015**, *27*, 273–287. [CrossRef]
22. Brandon, D.L.; Binder, R.G.; Bates, A.H.; Montague, W.C., Jr. Monoclonal antibody for multiresidue ELISA of benzimidazole anthelmintics in liver. *J. Agric. Food Chem.* **1994**, *42*, 1588–1594. [CrossRef]
23. Brandon, D.L.; Holland, K.P.; Dreas, J.S.; Henry, A.C. Rapid screening for benzimidazole residues in bovine liver. *J. Agric. Food Chem.* **1998**, *46*, 3653–3656. [CrossRef]
24. Eyer, K.; Doineau, R.C.L.; Castrillon, C.E.; Briseño-Roa, L.; Menrath, V.; Mottet, G.; England, P.; Godina, A.; Brient-Litzler, E.; Nizak, C.; et al. Single-cell deep phenotyping of IgG-secreting cells for high-resolution immune monitoring. *Nat. Biotechnol.* **2017**, *35*, 977–982. [CrossRef] [PubMed]
25. Mi, J.; Dong, X.; Zhang, X.; Li, C.; Wang, J.; Mujtaba, M.G.; Zhang, S.; Wen, K.; Yu, X.; Wang, Z. Novel hapten design, antibody recognition mechanism study, and a highly sensitive immunoassay for diethylstilbestrol in shrimp. *Anal. Bioanal. Chem.* **2019**, *411*, 5255–5265. [CrossRef]
26. Li, H.; Ma, S.; Zhang, X.; Li, C.; Dong, B.; Mujtaba, M.G.; Wei, Y.; Liang, X.; Yu, X.; Wen, K.; et al. Generic hapten synthesis, broad-specificity monoclonal antibodies preparation, and ultrasensitive ELISA for five antibacterial synergists in chicken and milk. *J. Agric. Food Chem.* **2018**, *66*, 11170–11179. [CrossRef] [PubMed]
27. Köhler, G.; Milstein, C. Continuous cultures of fused cells secreting antibody of predefined specificity. *Nature* **1975**, *256*, 495–497. [CrossRef] [PubMed]
28. Lu, T.; Chen, F. Multiwfn, A multifunctional wavefunction analyzer. *J Comput. Chem.* **2012**, *33*, 580–592. [CrossRef]
29. Lu, T.; Chen, F. Quantitative analysis of molecular surface based on improved marching tetrahedra algorithm. *J. Mol. Graph. Model.* **2012**, *38*, 314–323. [CrossRef]
30. Wen, K.; Bai, Y.; Wei, Y.; Li, C.; Shen, J.; Wang, Z. Influence of small molecular property on antibody response. *J. Agric. Food Chem.* **2020**, *68*, 10944–10950. [CrossRef]
31. Xu, Z.L.; Shen, Y.D.; Beier, R.C.; Yang, J.Y.; Lei, H.; Wang, H.; Sun, Y.M. Application of computer-assisted molecular modeling for immunoassay of low molecular weight food contaminants: A review. *Anal. Chim. Acta.* **2009**, *647*, 125–136. [CrossRef] [PubMed]
32. Li, C.; Liang, X.; Wen, K.; Li, Y.; Zhang, X.; Ma, M.; Yu, X.; Yu, W.; Shen, J.; Wang, Z. Class-specific monoclonal antibodies and dihydropteroate synthase in bioassays used for the detection of sulfonamides: Structural insights into recognition diversity. *Anal. Chem.* **2019**, *91*, 2392–2400. [CrossRef]
33. Wang, Z.; Zhu, Y.; Ding, S.; He, F.; Beier, R.C.; Li, J.; Jiang, H.; Feng, C.; Wan, Y.; Zhang, S.; et al. Development of a monoclonal antibody-based broad-specificity ELISA for fluoroquinolone antibiotics in foods and molecular modeling studies of cross-reactive compounds. *Anal. Chem.* **2007**, *79*, 4471–4483. [CrossRef]
34. Newman, D.J.; Price, C.P. Molecular aspects of design of immunoassays for drugs. *Ther. Drug Monit.* **1996**, *18*, 493–497. [CrossRef] [PubMed]
35. Wang, Z.; Luo, P.; Cheng, L.; Zhang, S.; Shen, J. Hapten-antibody recognition studies in competitive immunoassay of α-zearalanol analogs by computational chemistry and pearson correlation analysis. *J. Mol. Recognit.* **2011**, *24*, 815–823. [CrossRef] [PubMed]
36. Bai, Y.; Jiang, H.; Zhang, Y.; Dou, L.; Liu, M.; Yu, W.; Wen, K.; Shen, J.; Ke, Y.; Yu, X.; et al. Hydrophobic moiety of capsaicinoids haptens enhancing antibody performance in immunoassay: Evidence from computational chemistry and molecular recognition. *J. Agric. Food Chem.* **2021**, *69*, 9957–9967. [CrossRef] [PubMed]

foods

MDPI

Article

Development of a New Monoclonal Antibody against Brevetoxins in Oyster Samples Based on the Indirect Competitive Enzyme-Linked Immunosorbent Assay

Xiya Zhang [1], Mingyue Ding [1], Chensi Zhang [2], Yexuan Mao [1], Youyi Wang [1], Peipei Li [3], Haiyang Jiang [3], Zhanhui Wang [3] and Xuezhi Yu [3,*]

1 Henan Province Engineering Research Center for Food Safety Control of Processing and Circulation, College of Food Science and Technology, Henan Agricultural University, 63 Nongye Road, Zhengzhou 450002, China; zhangxiya@henau.edu.cn (X.Z.); dmy00117@163.com (M.D.); maoyexuan@henau.edu.cn (Y.M.); wangyouyi1213@163.com (Y.W.)

2 College of Life Sciences, Henan Agricultural University, 63 Nongye Road, Zhengzhou 450002, China; zhangchensi0719@163.com

3 Beijing Key Laboratory of Detection Technology for Animal–Derived Food Safety, Beijing Laboratory of Food Quality and Safety, College of Veterinary Medicine, China Agricultural University, Beijing 100193, China; 18331090835@163.com (P.L.); haiyang@cau.edu.cn (H.J.); wangzhanhui@cau.edu.cn (Z.W.)

* Correspondence: yu51422396@126.com; Tel.: +86-10-6273-4975

Citation: Zhang, X.; Ding, M.; Zhang, C.; Mao, Y.; Wang, Y.; Li, P.; Jiang, H.; Wang, Z.; Yu, X. Development of a New Monoclonal Antibody against Brevetoxins in Oyster Samples Based on the Indirect Competitive Enzyme-Linked Immunosorbent Assay. *Foods* **2021**, *10*, 2398. https://doi.org/10.3390/foods10102398

Academic Editor: Luís Abrunhosa

Received: 13 September 2021
Accepted: 2 October 2021
Published: 9 October 2021

Abstract: The consumption of shellfish contaminated with brevetoxins, a family of ladder-frame polyether toxins formed during blooms of the marine dinoflagellate *Karenia brevis*, can cause neurotoxic poisoning, leading to gastroenteritis and neurotoxic effects. To rapidly monitor brevetoxin levels in oysters, we generated a broad-spectrum antibody against brevetoxin 2 (PbTx-2), 1 (PbTx-1), and 3 (PbTx-3) and developed a rapid indirect competitive enzyme-linked immunosorbent assay (icELISA). PbTx-2 was reacted with carboxymethoxylamine hemihydrochloride (CMO) to generate a PbTx-2-CMO hapten and reacted with succinic anhydride (HS) to generate the PbTx-2-HS hapten. These haptens were conjugated to keyhole limpet hemocyanin (KLH) and bovine serum albumin (BSA) to prepare immunogen and coating antigen reagents, respectively, using the active ester method. After immunization and cell fusion, a broad-spectrum monoclonal antibody (mAb) termed mAb 1D3 was prepared. The 50% inhibitory concentration (IC_{50}) values of the icELISA for PbTx-2, PbTx-1, and PbTx-3 were 60.71, 52.61, and 51.83 μg/kg, respectively. Based on the broad-spectrum mAb 1D3, an icELISA was developed to determine brevetoxin levels. Using this approach, the limit of detection (LOD) for brevetoxin was 124.22 μg/kg and recoveries ranged between 89.08% and 115.00%, with a coefficient of variation below 4.25% in oyster samples. These results suggest that our icELISA is a useful tool for the rapid monitoring of brevetoxins in oyster samples.

Keywords: brevetoxins; monoclonal antibody; enzyme-linked immunosorbent assay

1. Introduction

Marine biotoxins have negative effects on the seafood industry. Typically, they are transferred along food chains and affect other organisms, including humans. Different types of poisoning are induced by marine biotoxins, e.g., puffer fish poisoning, paralytic shellfish poisoning, scombroid fish poisoning, diarrhetic shellfish poisoning, neurotoxic shellfish poisoning, ciguatera fish poisoning, and amnesic shellfish poisoning [1]. Brevetoxins belonging to the neurotoxic shellfish poisoning group are produced by the Florida red tide organism *Karenia brevis* and are divided in two groups: (1) those derived from the brevetoxin A backbone (PbTx-1, PbTx-7, and PbTx-10) and (2) those from brevetoxin B (PbTx-2, PbTx-3, PbTx-5, PbTx-6, PbTx-9, PbTx-11, and PbTx-12) [2]. PbTx-1 is the most potent, while PbTx-2 is the most highly produced brevetoxin (Figure 1) [3]. *K. brevis* blooms occur most years in the Gulf of Mexico, killing high numbers of fish and marine mammals, including sea turtles and aquatic birds, and generate economic losses of USD 2–24 million [4]. Physiologically, brevetoxins appear to activate voltage-sensitive sodium channels causing sodium influx and nerve membrane depolarization resulting in respiratory distress [5]. Thus, to protect human health and avoid food poisoning by brevetoxins, a rapid, sensitive, and specific assay for brevetoxins detection is required.

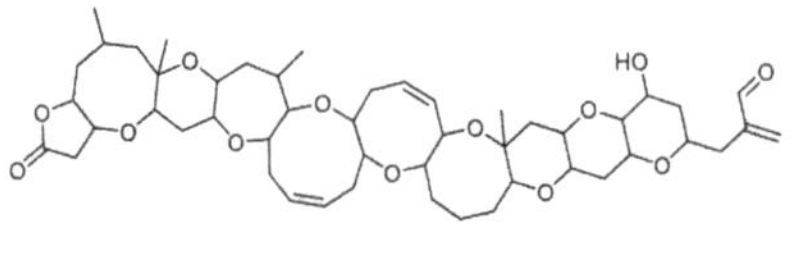

PbTx-1

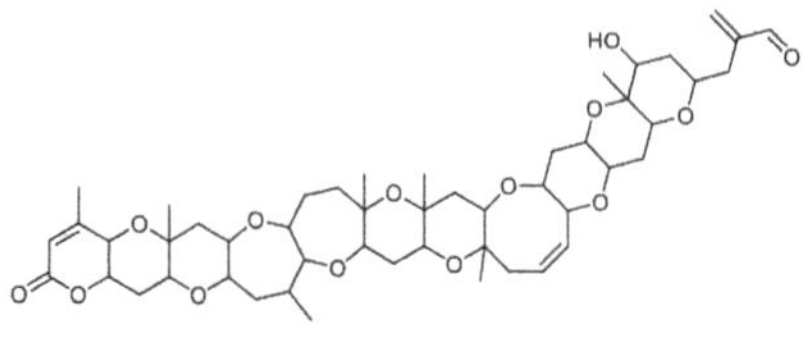

PbTx-2

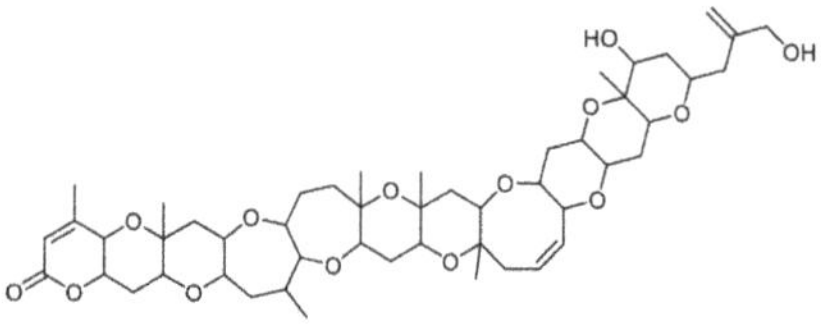

PbTx-3

Figure 1. PbTx-1, -2, and -3 structures.

Several analytical methods have been established to detect brevetoxins, including liquid chromatography–tandem mass spectrometry (LC-MS/MS) and receptor/antibody-based immunoassays. For example, Shin et al. (2018) developed an LC-MS/MS protocol for the PbTx-1, PbTx-2, and PbTx-3 brevetoxins with a limit of quantification (LOQ) of 25 μg/kg for each toxin [6]. Similarly, Wunschel et al. (2018) developed an electrospray LC-MS/MS system for the same brevetoxins with an LOQ of 2.5 μg/kg for each toxin [7]. Dom et al. (2018) established a high-resolution LC-MS system to detect 14 brevetoxins, with LOQs of 312 and 324 μg/kg for PbTx-2 in mussel and oyster, respectively [8]. However, these analytical methods are often time-consuming, expensive, and involve complex sample preparation. Therefore, new rapid brevetoxin screening/detection methods are required. McCall et al. (2012) developed a competitive binding assay based on rat brain synaptosomes as receptors and BODIPY®-PbTx-2 as competitive fluorescent probes for brevetoxin analogs [9]. In addition, Murata et al. (2019) developed a chemiluminescent receptor binding assay based on rat brain synaptosomes and acridinium-PbTx-2, with a detection limit value of 1.4 amol [10]. However, the synaptosomes were unstable and required storage at -80°C, and the assay was a time-consuming process.

Indirect competitive enzyme-linked immunosorbent assays (icELISAs) and lateral flow immunoassays (LFAs) are antibody-based and are frequently used as screening methods for small molecules due to their rapid turnaround, low costs, and high sensitivity. Recently, Zhou et al. (2010) prepared the monoclonal antibody (mAb) 2C4 using the PbTx-2-HS hapten; it exhibited IC_{50} values of 6.40, 6.57, and 5.31 μg/kg against PbTx-2, PbTx-1, and PbTx-3, respectively, with the accompanying icELISA having a limit of detection (LOD) of 0.60 ng/well [11]. Zhou et al. (2009) developed an LFA based on a colloidal gold probe for the rapid detection of brevetoxins in fish product samples, with a visual detection limit of 20 μg/kg [12]. Lai et al. (2016) developed a novel colorimetric immunoassay for PbTx-2 using an enzyme-controlled Fenton-based reaction and a 3,3′,5,5′-tetramethylbenzidine-based visual colored system, with an LOD of 0.08 ng/kg [13]. As is known, the preparation of broad-spectrum antibodies is a key step for developing an immunoassay. Hapten design is an important feature when preparing antibodies against target compounds. In the literature, PbTx-2 always reacted with succinic anhydride (HS) which introduced active sites as hapten [11,14]. In principle, aldehyde groups from PbTx-2 may be conjugated with a spacer arm, such as carboxymethoxylamine hemihydrochloride (CMO) and/or aminobenzoic acid. In addition, molecular modeling has become a powerful tool in guiding and improving hapten design strategies [15–19]. Therefore, we explored and developed novel haptens using molecular modeling to prepare mAbs against brevetoxins. We then developed an icELISA method to detect brevetoxins in oyster samples.

2. Materials and Methods

2.1. Materials and Reagents

PbTx-2, PbTx-1, PbTx-3, microcystins, nodularin (NOD), okadaic acid, keyhole limpet hemocyanin (KLH), bovine serum albumin (BSA), Freund's incomplete adjuvant (FIA), Freund's complete adjuvant (FCA), PEG1450, hypoxanthine aminopterin thymidine (HAT), and hypoxanthine thymidine (HT) were purchased from Sigma-Aldrich (St. Louis. MO, USA). Peroxidase-conjugated goat anti-mouse IgG was obtained from Jackson ImmunoResearch Laboratories, Inc. (West Grove, PA, USA). Succinic anhydride (HS), carboxymethoxylamine hemihydrochloride (CMO), N, N′-dicyclohexylcarbodiimide, N-hydroxysuccinimide, and tetramethylbenzidine were obtained from Aladdin Chemistry Co. Ltd. (Shanghai, China). Water was obtained from a Milli-Q purification system (Millipore Corp., Billerica, MA, USA). All other reagents were of analytical grade. Cell culture plates (24 and 96 wells) were from NEST (Wuxi, China). Polystyrene 96-well microtiter plates were from Costar Corp. (Cambridge, MA, USA). The absorbance at 450 nm was measured using a SpectraMax Mk3 microplate reader (Molecular Devices, Silicon Valley, CA, USA).

Coating buffer: Carbonate solution (0.05 mol/L, pH 9.60), 1.59 g Na_2CO_3, and 2.93 g $NaHCO_3$ were dissolved in 1 L water. Assay buffer: Phosphate-buffered saline (PBS, 0.01 mol/L, pH 7.4), 8.00 g NaCl, 0.20 g KCl, 2.93 g $NaHPO_4 \cdot 12H_2O$, and 0.20 g K_2PO_4 were dissolved in 1 L water. Washing buffer: The washing buffer comprised PBS containing 0.05% Tween 20 (PBST). Substrate solution: Solution A (pH 5) contained 1.00 g urea hydrogen peroxide, 18.00 g $Na_2HPO_4 \cdot 2H_2O$, and 10.30 g citric acid$\cdot H_2O$ per liter of water. Solution B (pH 2.40) contained 0.25 g tetramethylbenzidine, 40 mL N,N-dimethylformamide, 10.30 g citric acid$\cdot H_2O$, and 960 mL water. Before the assay, solutions A and B were mixed in a 1:1 ratio [20].

Eight-week-old female BALB/c mice were provided by Vital River Laboratory Animal Technology Co. Ltd. (Beijing, China) and raised under strictly controlled conditions. The mice were manipulated according to the China Agricultural University regulations concerning the protection of animals used for scientific purposes (2010-SYXK-0037).

2.2. Preparation of PbTx-2-Protein Conjugates

PbTx-2 was reacted with CMO to insert a carboxyl group to facilitate coupling to a carrier protein, as previously described but with modifications (Figure 2A) [21,22]. Briefly, 2 mg PbTx-2 in 5 mL pyridine was reacted with 2 mg CMO at 90°C for 6 h, after which the reaction mixture was evaporated to dryness under nitrogen gas. Then, the residue was dissolved in 5 mL 0.1 mol/L $NaHCO_3$ and extracted by 5 mL ethyl acetate. The aqueous phase was adjusted to pH 3 with 0.05 mol/L HCl and extracted three times in 5 mL ethyl acetate. The organic phase was dried under N_2 at 40 °C to generate the PbTx-2-CMO hapten. The PbTx-2-HS hapten was similarly synthesized by reacting PbTx-2 with HS as described (Figure 2B).

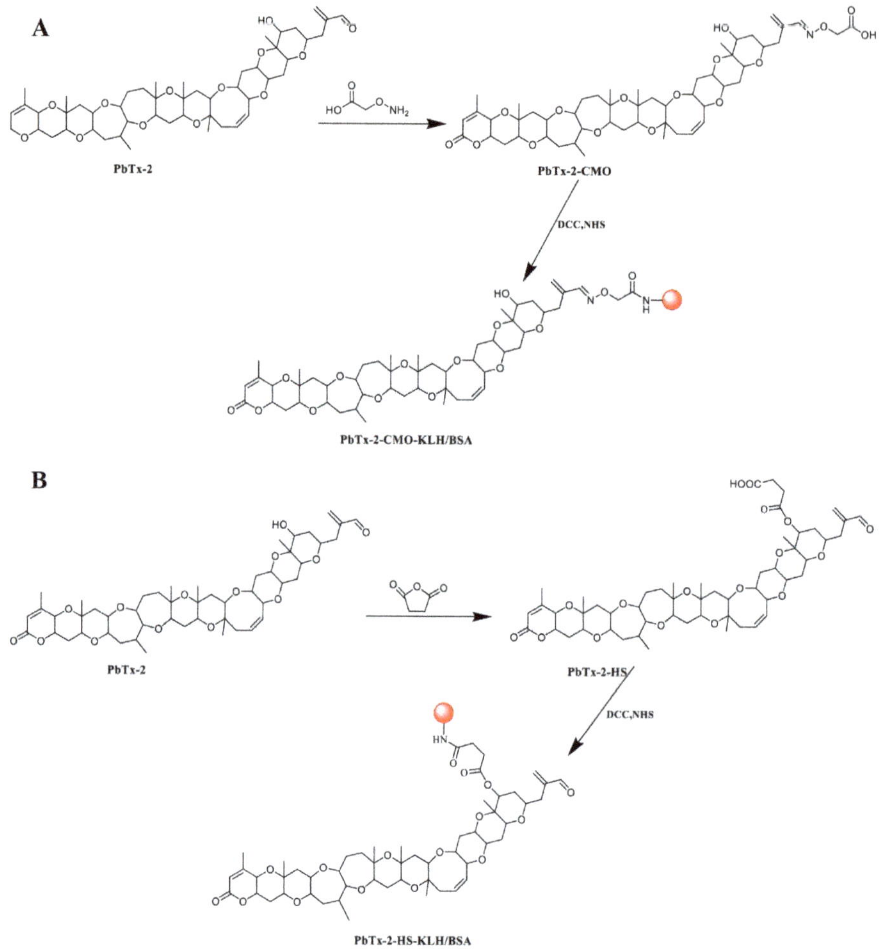

Figure 2. PbTx-2 hapten and antigen synthesis routes. (**A**) PbTx-2-CMO-KLH/BSA. (**B**) PbTx-2HS-KLH/BSA.

Then, the PbTx-2-CMO and PbTx-2-HS haptens were conjugated to KLH (immunogen) and BSA (coating antigen), respectively, via the active ester method [23]. Briefly, the haptens were respectively redissolved in 0.5 mL N,N-dimethylformamide containing 2 mg N,N'-dicyclohexylcarbodiimide and 1.5 mg N-hydroxysuccinimide and reacted for 12 h at room temperature. After centrifugation at 8000× *g* for 10 min, the supernatant of each active hapten solution was divided into two and added drop-wise to 4 mL PBS containing 5 mg KLH and 10 mL PBS containing 10 mg BSA, respectively. Reaction mixtures were stirred at 4 °C for 12 h, and the conjugates of PbTx-2-CMO-KLH/BSA and PbTx-2-HS-KLH/BSA were dialyzed in PBS for 48 h. Antigens were then stored at −20 °C.

2.3. Preparation of mAbs

Animal immunization procedures were as follows: twelve female BALB/c mice were immunized by subcutaneous injection. The immunogens PbTx-2-CMO-KLH and PbTx-2-HS-KLH (100 μg) were emulsified in Freund's complete adjuvant for the first immunizations. After this, the mice were boosted with the same immunogen doses in Freund's incomplete adjuvant every 3 weeks. Then, 7–10 days after the last immunization, serum was collected from the caudal vein. Antibody titers were analyzed by ELISA and specificity was characterized by icELISA. Animals with the highest inhibition ratios were sacrificed for fusion studies [23]. The inhibition ratio was calculated as follows:

$$\text{Inhibition ratio (\%)} = (1 - B/B_0) \times 100\%, \quad (1)$$

where B_0 and B correspond to the absorbance value of wells without a standard and the absorbance value of wells with *x* μg/kg PbTx-2 standard, respectively. Spleen cells from animals with the highest inhibition ratios were collected and fused with Sp2/0 myeloma cells using PEG1450 at a 10:1 ratio. Fusion cells were cultured in hypoxanthine aminopterin thymidine medium for 7 days. Supernatants from fusion cultures were also assayed for the titer and the inhibition ratios using ELISA and icELISA. Positive hybridomas were subcloned three times using the limiting dilution method and injected into BALB/c mice to produce ascites [23].

2.4. ELISA and icELISA Protocols

ELISA was conducted as previously described [24]. Briefly, (1) 100 μL of coating antigen PbTx-2-CMO-BSA (or PbTx-2-HS-BSA) diluted in coating buffer at 1 μg/mL was added to the wells of a 96-well plate and incubated at 4 °C for 10–12 h. (2) The coating antigen was then discarded, and the plate was washed three times in wash buffer (250 μL/well). (3) Then, 200 μL of 2% skimmed milk powder in assay buffer was added per well and incubated at 37 °C for 1 h to reduce unspecific binding. (4) After this, 50 μL 10 mM PBS and 50 μL anti-PbTx-2 mAb diluted in assay buffer were added sequentially to wells and incubated at 37 °C for 30 min. (5) After further washing, 100 μL/well peroxidase-conjugated goat anti-mouse IgG (diluted 1:5000) was added and incubated at 37 °C for 30 min. (6) The plate was washed five times, 100 μL/well substrate solution was added to the reaction, and the plate was incubated at 37 °C for 15 min. (7) The reaction was stopped with 50 μL 2 M H_2SO_4, and the absorbance was detected at 450 nm on a Multiskan MK3 microplate reader.

For the icELISA procedure, the 50 μL 10 mM PBS was replaced by 50 μL series concentration of PbTx-2 standard solution at step 4.

We characterized mAb properties using IC_{50} and cross-reactivity (CR) values. Sensitivity was assessed using IC_{50} values, with the concentration of the competitor resulting in the inhibition ratio reaching 50%, and specificity was evaluated by CR based on the following formula [24]:

$$\text{CR (\%)} = (IC_{50} \text{ of PbTx-2}/IC_{50} \text{ of competitors}) \times 100\% \quad (2)$$

We selected several marine toxin compounds as competitors, including PbTx-1 (100 μg/kg), PbTx-3 (100 μg/kg), domoic acid (1000 μg/kg), microcystins (1000 μg/kg), nodularin (1000 μg/kg), neosaxitoxin (1000 μg/kg), and tetrodotoxin (1000 μg/kg).

2.5. Sample Preparation

Blank oyster samples from a local supermarket were confirmed using the HPLC-MS/MS method [6]. The blank oyster samples were spiked with 200, 400, and 800 μg/kg. Then, brevetoxins were extracted from the oysters based on a previously described method with some modifications [6]. Briefly, 5 g homogenized sample was extracted in 5 mL 80% methanol by vortexing for 5 min and then ultrasonication at 60 °C for 10 min. After centrifugation at 2400× *g* for 10 min, the supernatant was diluted 15-fold in assay buffer to eliminate matrix interference. The LOD value in oyster samples was calculated based on 20 blank samples, accepting no false positive rates, with an average value plus triple standard deviation (SD), then multiplied by 15 (the diluted ratio) [23]. The recovery was calculated as the following equation [23]:

$$\text{Recovery (\%)} = (\text{measured value } (\mu g/kg)/\text{spiked values } (\mu g/kg)) \times 100\% \quad (3)$$

2.6. Molecule Alignment and Electrostatic Potential Analysis

PbTx-2, PbTx-2-CMO, and PbTx-2-HS structures were constructed using Gaussian 09 software (Gaussian Wallingford, CT, USA) according to PbTx-2 configurations in the PubChem database [15]. PbTx-2 was selected as the template to align PbTx-2-CMO and PbTx-2-HS haptens using Discovery Studio 2016 software (Accelrys Software, Inc., San Diego, CA, USA). Both Gaussian 09 and Gaussian View 5 packages were used to conduct molecular electrostatic potential (ESP) analysis.

3. Results and Discussion

3.1. Hapten Design, Synthesis, and Conjugate Preparation

Hapten design is key to generating excellent antibody performances against small molecules [25]. Generally, haptens should mimic the target molecule in terms of size, shape, electronic properties, and insert functional groups such as, carboxyl, amino, and sulfhydryl groups for carrier protein coupling [16,17]. In principle, there are two ways to synthesize PbTx-2 haptens. First, the hydroxy group of PbTx-2 is reacted with HS, which introduces active sites to construct the PbTx-2-HS hapten (Figure 2A) [11,14]. Moreover, the PbTx-2 aldehyde group is an active site which potentially reacts with an amino group with the objective of obtaining a probe or hapten. For example, McCall et al. (2012) used the active site of the aldehyde group of PbTx-2 to couple with BODIPY® to synthesize the BODIPY®-PbTx-2 fluorescence probe [9]. Murata et al. (2019) also used this site to prepare the acridinium-PbTx-2 fluorescence probe [10]. Thus, PbTx-2 was reacted with CMO to prepare the hapten of PbTx-2-CMO (Figure 2B).

To further design the optimal hapten, PbTx-2 was selected as the template molecule to align the haptens of PbTx-2-CMO and PbTx-2-HS based on their lowest energy conformation. As shown in Figure 3A, the PbTx-2-CMO and PbTx-2-HS haptens were exposed on the left side of the structure to animal immunity. In addition, the introduction of spacer arms at the O57 or O61 position of PbTx-2 barely influenced the atom partial charges feature of PbTx-2 (Figure 3B). However, the introduction of HS hapten caused the spacer arm to form a certain angle with the parent structure of PbTx-2 (Figure 3A). To further explain the difference between PbTx-2-CMO, PbTx-2-HS, and the target compound PbTx-2 in terms of conformation and electron distribution, the ESP displayed on the van der Waals surfaces of global lowest energy conformation for PbTx-2, PbTx-2-CMO, and PbTx-2-HS is shown (Figure 3C). The PbTx-2-CMO conformation had the most similar structure to the target, PbTx-2 (Figure 3C, points A and B). The spacer HS arm of PbTx-2-HS formed a specific spatial conformation with the parent nucleus structure (Figure 3C, point C), non-conducive to the production of high-affinity antibodies against the target. Thus, PbTx-2-CMO was the ideal hapten to be used for antibody production against PbTx-2. To further verify the quality of haptens, PbTx-2-CMO and PbTx-2-HS were synthesized and conjugated with the carrier protein as complete antigens.

PbTx-2-CMO and PbTx-2-HS were activated by DCC and NHS and then conjugated with KLH and BSA as an immunogen and a coating antigen, respectively. The UV–visible absorption spectra of the immunogens, PbTx-2-CMO-KLH and PbTx-2-HS-KLH, and the coating antigens, PbTx-2-CMO-BSA and PbTx-2-HS-BSA, are shown (Figure 4). PbTx-2 had an absorbance peak at 264 nm, KLH at 280 and 350 nm, and BSA at 279 nm. The maximum absorbance peaks for PbTx-2-CMO-KLH, PbTx-2-HS-KLH, PbTx-2-CMO-BSA, and PbTx-2-HS-BSA were 275, 277, 274, and 278 nm, respectively. The maximum absorbance peaks between conjugates and PbTx-2 had shifted, thus indicating the successful synthesis of complete antigens. The calculated molar ratio of hapten to carrier protein was 1.5 for PbTx-2-CMO-KLH, 0.9 for PbTx-2-HS-KLH, 1.2 for PbTx-2-CMO-BSA, and 0.6 for PbTx-2-HS-BSA. It is accepted that the coupling ratio is an important factor affecting generated antibodies; higher coupling ratios could induce higher antibody titers [17]. In the literature, appropriate coupling ratios ranged from 3 to 15 [16,17,21]. The coupling ratios of PbTx-2-CMO-KLH/BSA and PbTx-2-HS-KLH/BSA were relatively low due to the lower reaction molar ratio (hapten to carrier protein). However, the conjugation of PbTx-2-CMO-

KLH/BSA and PbTx-2-HS-KLH/BSA was successful according to the UV–visible absorption spectra (Figure 4). Therefore, these antigens were used for immunization studies.

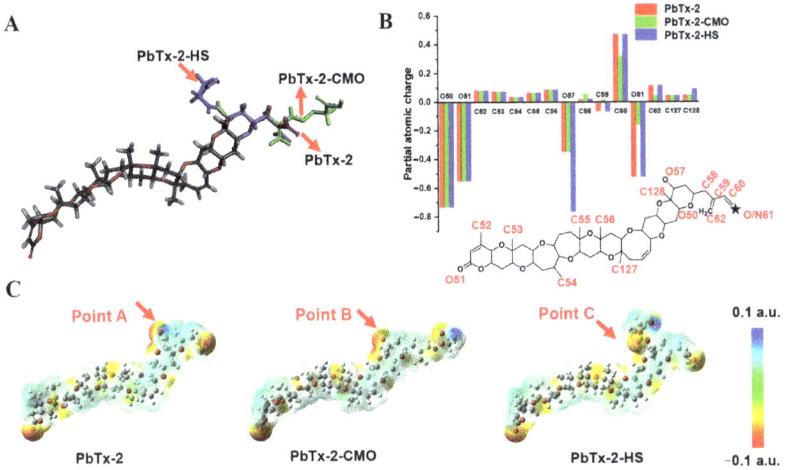

Figure 3. Molecular modeling results. (**A**) Overlap of PbTx-2 (gray), PbTx-2-CMO (green), and PbTx-2-HS (violet) structures. (**B**) Calculated partial atomic charges of PbTx-2, PbTx-2-CMO, and PbTx-2-HS structures. (**C**) ESP for PbTx-2, PbTx-2-CMO, and PbTx-2-HS structures. Red and blue areas indicate negative and positive potentials, respectively.

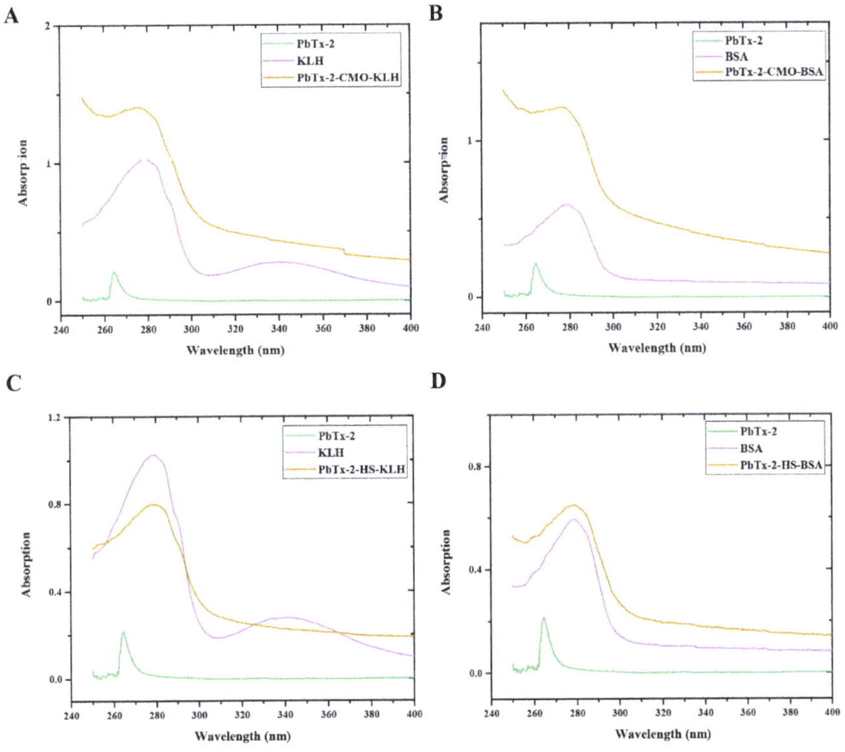

Figure 4. UV–visible absorption spectra of PbTx-2-CMO-KLH (**A**), PbTx-2-CMO-BSA (**B**), PbTx-2-CMO-KLH (**C**) and PbTx-2-CMO-BSA (**D**).

3.2. Antibody Production and Characterization

The two immunogens PbTx-2-CMO-KLH and PbTx-2-HS-KLH were used to generate antibodies against PbTx-2. After a third immunization, serum was collected and characterized using ELISA and icELISA (Table 1). Mice immunized with both complete antigens produced antiserum against PbTX-2; the immune response to PbTx-2-CMO-KLH was superior to that to PbTx-2-HS-KLH from an inhibition rate perspective. However, the PbTx-2-HS-KLH antibody titer was higher than that for PbTx-2-CMO-KLH, possibly suggesting the spacer HS arm of the PbTx-2-HS molecule that formed based on molecular modeling (Figure 3A,C), a specific spatial conformation with the parent nucleus structure conduced to the production of high-titer antibodies. Additionally, all the antiserum titers from mice were low due to the lower coupling ratios of the hapten to the carrier protein. Finally, mouse No. 1 (PbTx-2-CMO-KLH) and mouse No. 5 (PbTx-2-HS-KLH) were sacrificed for fusion studies as they exhibited a higher inhibition ratio and antiserum titer.

Table 1. Immunization with PbTx-2-CMO-KLH and PbTx-2-HS-KLH.

Immunogen: PbTx-2-CMO-KLH/Coating Antigen PbTx-2-CMO-BSA [a]						
No.	**1**	**2**	**3**	**4**	**5**	**6**
Absorbance of PbTx at 0 mg/kg	1.64 ± 0.06	0.52 ± 0.07	0.74 ± 0.07	0.76 ± 0.08	0.98 ± 0.07	1.542 ± 0.06
Absorbance of PbTx at 0.10 mg/kg [c]	0.51 ± 0.02	0.15 ± 0.01	0.17 ± 0.01	0.20 ± 0.03	0.29 ± 0.05	0.63 ± 0.06
Absorbance of PbTx at 1.00 mg/kg [c]	0.11 ± 0.02	0.07 ± 0.01	0.08 ± 0.02	0.09 ± 0.01	0.11 ± 0.01	0.14 ± 0.01
Inhibition ratio by 0.10 mg/kg	68.90%	71.15%	77.03%	84.80%	70.41%	59.14%
Inhibition ratio by 1.00 mg/kg	93.29%	86.54%	89.19%	88.16%	88.78%	90.92%
Immunogen: PbTx-2-HS-KLH/Coating Antigen PbTx-2-HS-BSA [b]						
No.	**1**	**2**	**3**	**4**	**5**	**6**
Absorbance of PbTx at 0 mg/L	2.39 ± 0.05	0.71 ± 0.04	0.28 ± 0.01	1.07 ± 0.08	1.68 ± 0.14	0.79 ± 0.01
Absorbance of PbTx at 0.50 mg/L [c]	2.13 ± 0.07	0.63 ± 0.05	0.17 ± 0.01	0.59 ± 0.01	1.14 ± 0.07	0.69 ± 0.04
Absorbance of PbTx at 5.00 mg/L [c]	1.77 ± 0.03	0.58 ± 0.07	0.16 ± 0.05	0.34 ± 0.06	0.96 ± 0.06	0.49 ± 0.01
Inhibition ratio by 0.50 mg/kg	10.88%	14.86%	39.29%	44.86%	32.14%	12.66%
Inhibition ratio by 5.00 mg/kg	25.94%	18.31%	42.86%	68.22%	42.86%	37.97%

[a] The concentration of coating antigen was 1.00 μg/mL, and the antibody titer was 1:200. [b] The concentration of coating antigen was 0.50 μg/mL, and the antibody titer was 1:1000. [c] The competition compound was PbTx-2.

After cell fusion, the 1D3 cell line from mouse No.1 (PbTx-2-CMO-KLH) and the cell lines 6D8 and 9E8 from mouse No. 5 (PbTx-2-HS-KLH) were identified as secreting antibodies against PbTx-2. Thus, all three cell lines were used for antibody production, and ELISAs and icELISAs were used to characterize mAbs (Table 2). All mAbs recognized PbTx-2, PbTx-1, and PbTx-3. Additionally, the IC_{50} value for the 1D3 mAb was lower than that for the 6D8 and 9G8 mAbs, consistent with the molecular modeling data (Figure 3). Those results also indicated that PbTx-2-CMO was the best hapten for brevetoxin production, and the PbTx-2-HS-BSA coating antigen improved 1D3 mAb sensitivity.

Table 2. IC_{50} values (μg/kg) and cross-reaction (CR) of monoclonal antibodies.

Compound	**PbTx-2**	**CR (%)**	**PbTx-1**	**CR (%)**	**PbTx-3**	**CR (%)**
mAb 6D8 [a]	786.72	100	806.52	97.92	726.51	108.29
mAb 9G8 [a]	434.54	100	424.61	102.33	398.62	109.01
mAb 1D3 [b]	78.53	100	80.62	97.41	69.84	112.44
mAb 1D3 [c]	60.71	100	52.61	115.40	51.83	117.13

[a] The concentration of coating antigen PbTx-2-HS-BSA was 0.20 μg/mL, and the antibody titer was $1{:}1 \times 10^5$.
[b] The concentration of coating antigen PbTx-2-CMO-BSA was 1.00 μg/mL, and the antibody titer was $1{:}\ 1 \times 10^4$.
[c] The concentration of coating antigen PbTx-2-HS-BSA was 0.20 μg/mL, and the antibody titer was $1{:}\ 3 \times 10^4$.

Next, PbTx-2-HS-BSA and 1D3 mAb were used to establish a standard curve in buffer assay. As shown (Figure 5), the curves that were based on the PbTx-2, PbTx-1 and PbTx-3 gave IC_{50} values of 60.71 μg/kg, 52.61 μg/kg and 51.83 μg/kg, respectively. The mAb 1D3 exhibited a CR = 100% toward PbTx-2, CR = 115.40% toward PbTx-1, and CR = 117.13% toward PbTx-3. The LOD in assay buffer was 6.11 μg/kg, and the linearity range (IC_{10}–IC_{90}) was between 7.46 and 127.00 μg/kg. In addition, the mAb did not exhibit a measurable CR with other marine biotoxins, including domoic acid, microcystins, nodularin, neosaxitoxin, and tetrodotoxin. These data indicated that mAb 1D3 was a broad spectrum homogeneous antibody for brevetoxins.. The sensitivity of the mAb 1D3 icELISA was lower than that for the mAb 2C4 (IC_{50} value of 5.3 μg/L towards PbTx-2) [11]. In this study, the lower coupling ratios of PbTx-2-CMO to the carrier protein was because of the lower feed ratios (2 mg PbTx-2 reacted with 2 mg CMO), resulting in lower antibody titer immunized by PbTx-2-CMO-KLH. Ultimately, this affected the performance parameters of the mAb 1D3. More sensitive antibodies can be obtained if the ratios of haptens (PbTx-2-CMO) to carrier proteins are improved.

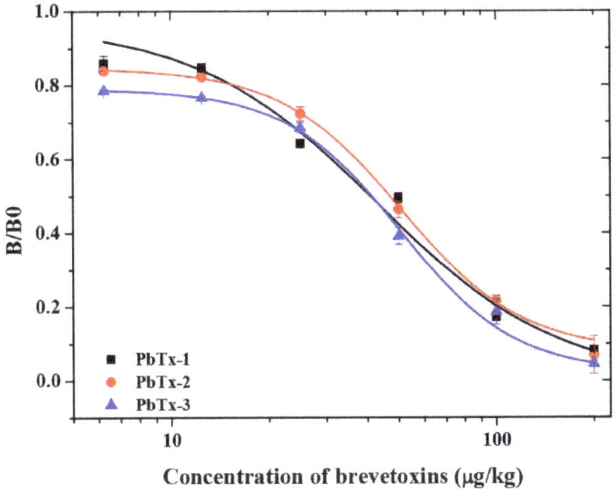

Figure 5. The brevetoxin icELISA standard curve using the mAb 1D3 with the PbTx-2-HS-BSA coating antigen.

The recoveries for PbTx-2, PbTx-1, and PbTx-3 from spiked oyster samples at three dose levels (200, 400, and 800 μg/kg) are shown in Table 3 and were 91.21–108.33%, 91.04–115.00%, and 89.08–113.17%, respectively. The LOD in the oyster sample was calculated at 124.22 μg/kg. The sensitivity of this icELISA was higher than the high-resolution LC-MS sensitivity (LOQ of 324 μg/kg for PbTx-2 in oyster samples) [8] and similar to that of the LFA based on colloidal gold probe (visual detection limit of 20 μg/kg in fish product samples) by Zhou et al. [12].

Table 3. Toxin recoveries from spiked oyster samples.

Toxin	Concentration (μg/kg)	Recovery of Toxins from Oyster
	200	108.33 ± 3.82%
PbTx-2	400	91.25 ± 3.31%
	800	91.21 ± 1.25%
	200	115.00 ± 5.00%
PbTx-1	400	92.92 ± 3.82%
	800	91.04 ± 3.21%
	200	113.17 ± 4.25%
PbTx-3	400	94.42 ± 3.12%
	800	89.08 ± 1.12%

4. Conclusions

Using molecular modeling and experimental analyses, the PbTx-2-CMO hapten produced acceptable antibody characteristics against brevetoxins. The IC_{50} values against PbTx-2, PbTx-1, and PbTx-3 were 60.71, 52.61, and 51.83 μg/kg, respectively. The LOD was 124.22 μg/kg, and PbTx recoveries from oysters ranged from 89.08% to 115.00%. This icELISA will be a useful method for monitoring PbTxs in oyster samples.

Author Contributions: Conceptualization, X.Z.; methodology, X.Z. and M.D.; software, M.D.; validation, C.Z.; formal analysis, Y.M.; data curation, Y.W. and P.L.; writing—original draft preparation, X.Z. and H.J.; writing—review and editing, Z.W. and X.Y.; visualization, X.Z. and M.D.; supervision, X.Y.; project administration, X.Y.; funding acquisition, X.Y. All authors have read and agreed to the published version of the manuscript.

Funding: This work was supported by the Chinese Ministry of Science and Technology (MOST) for the National Key R&D Program of China (2018YFC1602600), the Natural Science Foundation of China (No. 31802249), the Science and Technology Key Research Project of Henan Provincial Education Department of China (No. 20A550009), and the Key Scientific and Technological Project of Henan Province Department of China (No. 202102310303).

Institutional Review Board Statement: Not applicable.

Informed Consent Statement: Not applicable.

Data Availability Statement: Not applicable.

Conflicts of Interest: The authors declare no conflict of interest.

References

1. Özogul, F.; Hamed, I. Marine-based toxins and their health risk. In *Food Quality: Balancing Health and Disease*; Academic Press: Cambridge, MA, USA, 2018; pp. 109–144.
2. Radwan, F.F.Y.; Wang, Z.H.; Ramsdell, J.S. Identification of a rapid detoxification mechanism for brevetoxin in rats. *Toxicol. Sci.* **2005**, *85*, 839–846. [CrossRef]
3. Landsberg, J.H. The effects of harmful algal blooms on aquatic organisms. *Rev. Fish. Sci.* **2002**, *10*, 113–390.
4. Hoagland, P.; Jin, D.; Beet, A.; Kirkpatrick, B.; Reich, A.; Ullmann, S.; Fleming, L.E.; Kirkpatrick, G. The human health effects of Florida red tide (FRT) blooms: An expanded analysis. *Environ. Int.* **2014**, *68*, 144–153. [CrossRef] [PubMed]
5. Baden, D.G.; Bourdelais, A.J.; Jacocks, H.; Michelliza, S.; Naar, J. Natural and derivative brevetoxins: Historical background, multiplicity, and effects. *Environ. Health Perspect.* **2005**, *113*, 621–625. [CrossRef] [PubMed]
6. Shin, C.; Hwang, J.Y.; Yoon, J.H.; Kim, S.H.; Kang, G.J. Simultaneous determination of neurotoxic shellfish toxins (brevetoxins) in commercial shellfish by liquid chromatography tandem mass spectrometry. *Food Control.* **2018**, *91*, 365–371. [CrossRef]
7. Wunschel, D.S.; Valenzuela, B.R.; Kaiser, B.L.D.; Victry, K.; Woodruff, D. Method development for comprehensive extraction and analysis of marine toxins: Liquid-liquid extraction and tandem liquid chromatography separations coupled to electrospray tandem mass spectrometry. *Talanta* **2018**, *187*, 302–307. [CrossRef]
8. Dom, I.; Bire, R.; Hort, V.; Lavison-Bompard, G.; Nicolas, M.; Guerin, T. Extended targeted and non-targeted strategies for the analysis of marine toxins in mussels and oysters by (LC-HRMS). *Toxins* **2018**, *10*, 375. [CrossRef]
9. McCall, J.R.; Jacocks, H.M.; Baden, D.G.; Bourdelais, A.J. Development of a competitive fluorescence-based synaptosome binding assay for brevetoxins. *Harmful Algae* **2012**, *19*, 85–91. [CrossRef]
10. Murata, K.; Yasumoto, T. Chemiluminescent receptor binding assay for ciguatoxins and brevetoxins using acridinium brevetoxin-b2. *Toxins* **2019**, *11*, 580. [CrossRef]
11. Zhou, Y.; Li, Y.S.; Pan, F.G.; Zhang, Y.Y.; Lu, S.Y.; Ren, H.L.; Li, Z.H.; Liu, Z.S.; Zhang, J.H. Development of a new monoclonal antibody based direct competitive enzyme-linked immunosorbent assay for detection of brevetoxins in food samples. *Food Chem.* **2010**, *118*, 467–471. [CrossRef]
12. Zhou, Y.; Pan, F.G.; Li, Y.S.; Zhang, Y.Y.; Zhang, J.H.; Lu, S.Y.; Ren, H.L.; Liu, Z.S. Colloidal gold probe-based immunochromatographic assay for the rapid detection of brevetoxins in fishery product samples. *Biosens. Bioelectron.* **2009**, *24*, 2744–2747. [CrossRef]
13. Lai, W.; Wei, Q.; Zhuang, J.; Lu, M.; Tang, D. Fenton reaction-based colorimetric immunoassay for sensitive detection of brevetoxin B. *Biosens. Bioelectron.* **2016**, *80*, 249–256. [CrossRef]
14. Naar, J.; Branaa, P.; Bottein-Dechraoui, M.Y.; Chinain, M.; Pauillac, S. Polyclonal and monoclonal antibodies to PbTx-2-type brevetoxins using minute amount of hapten-protein conjugates obtained in a reversed micellar medium. *Toxicon* **2001**, *39*, 869–878. [CrossRef]

15. Li, H.; Ma, S.; Zhang, X.; Li, C.; Dong, B.; Mujtaba, M.G.; Wei, Y.; Liang, X.; Yu, X.; Wen, K.; et al. Generic hapten synthesis, broad-specificity monoclonal antibodies preparation, and ultrasensitive elisa for five antibacterial synergists in chicken and milk. *J. Agric. Food Chem.* **2018**, *66*, 11170–11179. [CrossRef] [PubMed]
16. Mari, G.M.; Li, H.; Dong, B.; Yang, H.; Talpur, A.; Mi, J.; Guo, L.; Yu, X.; Ke, Y.; Han, D.; et al. Hapten synthesis, monoclonal antibody production and immunoassay development for direct detection of 4-hydroxybenzehydrazide in chicken, the metabolite of nifuroxazide. *Food Chem.* **2021**, *355*, 129598. [CrossRef]
17. Li, Z.; Wang, Y.; Li, D.; Chen, X.; Li, Z.; Gao, H.; Cao, L.; Li, S.; Hou, Y. Development of an indirect competitive enzyme-linked immunosorbent assay for screening ethopabate residue in chicken muscle and liver. *RSC Adv.* **2017**, *7*, 36072–36080. [CrossRef]
18. Han, X.; Sheng, F.; Kong, D.; Wang, Y.; Pan, Y.; Chen, M.; Tao, Y.; Liu, Z.; Ahmed, S.; Yuan, Z.; et al. Broad-spectrum monoclonal antibody and a sensitive multi-residue indirect competitive enzyme-linked immunosorbent assay for the antibacterial synergists in samples of animal origin. *Food Chem.* **2019**, *280*, 20–26. [CrossRef]
19. Huang, J.X.; Yao, C.Y.; Yang, J.Y.; Li, Z.F.; He, F.; Tian, Y.X.; Wang, H.; Xu, Z.L.; Shen, Y.D. Design of novel haptens and development of monoclonal antibody-based immunoassays for the simultaneous detection of tylosin and tilmicosin in milk and water samples. *Biomolecules* **2019**, *9*, 770. [CrossRef] [PubMed]
20. O'Keeffe, M.; Crabbe, P.; Salden, M.; Wichers, J.; Van Peteghem, C.; Kohen, F.; Pieraccini, G.; Moneti, G. Preliminary evaluation of a lateral flow immunoassay device for screening urine samples for the presence of sulphamethazine. *J. Immunol. Methods* **2003**, *278*, 117–126. [CrossRef]
21. Zhang, X.; Wen, K.; Wang, Z.; Jiang, H.; Beier, R.C.; Shen, J. An ultra-sensitive monoclonal antibody-based fluorescent microsphere immunochromatographic test strip assay for detecting aflatoxin M_1 in milk. *Food Control* **2016**, *60*, 588–595. [CrossRef]
22. Zhang, X.; Eremin, S.A.; Wen, K.; Yu, X.; Li, C.; Ke, Y.; Jiang, H.; Shen, J.; Wang, Z. Fluorescence polarization immunoassay based on a new monoclonal antibody for the detection of the zearalenone class of mycotoxins in maize. *J. Agric. Food Chem.* **2017**, *65*, 2240–2247. [CrossRef] [PubMed]
23. Zhang, X.; Song, M.; Yu, X.; Wang, Z.; Ke, Y.; Jiang, H.; Li, J.; Shen, J.; Wen, K. Development of a new broad-specific monoclonal antibody with uniform affinity for aflatoxins and magnetic beads-based enzymatic immunoassay. *Food Control.* **2017**, *79*, 309–316. [CrossRef]
24. Peng, D.; Chang, F.; Wang, Y.; Chen, D.; Liu, Z.; Zhou, X.; Feng, L.; Yuan, Z. Development of a sensitive monoclonal-based enzyme-linked immunosorbent assay for monitoring T-2 toxin in food and feed. *Food Addit. Contam. Part A Chem. Anal. Control. Expo. Risk Assess.* **2016**, *33*, 683–692. [CrossRef]
25. Wang, Z.; Liu, M.; Shi, W.; Li, C.; Zhang, S.; Shen, J. New haptens and antibodies for ractopamine. *Food Chem.* **2015**, *183*, 111–114. [CrossRef] [PubMed]

foods

Article

Preparation and Directed Evolution of Anti-Ciprofloxacin ScFv for Immunoassay in Animal-Derived Food

Fangyu Wang [1,*,†], Ning Li [2,†], Yunshang Zhang [1], Xuefeng Sun [1], Man Hu [1], Yali Zhao [2] and Jianming Fan [3]

[1] Key Laboratory for Animal Immunology, Henan Academy of Agricultural Sciences, 116#Huayuan Road, Zhengzhou 450002, China; yunshangzh@163.com (Y.Z.); sunxuefeng2021@126.com (X.S.); human131@163.com (M.H.)
[2] Department of Food Nutrition and Health, College of Food Science and Technology, Henan Agricultural University, 63#Agricultural Road, Zhengzhou 450000, China; ln8028@163.com (N.L.); zhaoyali9016@163.com (Y.Z.)
[3] China College of Public Health, Zhengzhou University, 100#Kexue Avenue, Zhengzhou 450001, China; fan5746067@126.com
* Correspondence: sprinkle.w@126.com
† Both authors contributed equally to this work.

Abstract: An immunized mouse phage display scFv library with a capacity of 3.34×10^9 CFU/mL was constructed and used for screening of recombinant anti-ciprofloxacin single-chain antibody for the detection of ciprofloxacin (CIP) in animal-derived food. After four rounds of bio-panning, 25 positives were isolated and identified successfully. The highest positive scFv-22 was expressed in *E. coli* BL21. Then, its recognition mechanisms were studied using the molecular docking method. The result showed the amino acid residue Val160 was the key residue for the binding of scFv to CIP. Based on the results of virtual mutation, the scFv antibody was evolved by directional mutagenesis of contact amino acid residue Val160 to Ser. After the expression and purification, an indirect competitive enzyme-linked immunosorbent assay (IC-ELISA) based on the parental and mutant scFv was established for CIP, respectively. The IC50 value of the assay established with the ScFv mutant was 1.58 ng/mL, while the parental scFv was 26.23 ng/mL; this result showed highly increased affinity, with up to 16.6-fold improved sensitivity. The mean recovery for CIP ranged from 73.80% to 123.35%, with 10.46% relative standard deviation between the intra-assay and the inter-assay. The RSD values ranged between 1.49% and 9.81%. The results indicate that we obtained a highly sensitive anti-CIP scFv by the phage library construction and directional evolution, and the scFv-based IC-ELISA is suitable for the detection of CIP residue in animal-derived edible tissues.

Keywords: scFv; ciprofloxacin; recognition mechanism; directional mutagenesis; IC-ELISA

Citation: Wang, F.; Li, N.; Zhang, Y.; Sun, X.; Hu, M.; Zhao, Y.; Fan, J. Preparation and Directed Evolution of Anti-Ciprofloxacin ScFv for Immunoassay in Animal-Derived Food. *Foods* **2021**, *10*, 1933. https://doi.org/10.3390/foods10081933

Academic Editor: Thierry Noguer

Received: 23 June 2021
Accepted: 17 August 2021
Published: 20 August 2021

1. Introduction

Ciprofloxacin (CIP) is a synthetic third-generation fluoroquinolone (FQ) antibiotic that has been developed and is widely used to treat bacterial infections in humans and animals. This antibiotic exerts effects by inhibiting DNA gyrase or topoisomerase II in susceptible bacteria and exhibits high activity against a broad spectrum of Gram-negative and Gram-positive bacteria [1]. However, the unreasonable and extensive use of antibiotics has resulted in the potential for residual antibiotics in food of animal origin, which can damage multiple systems in the body [2,3] and cause bacterial resistance [4,5]. Therefore, the European Union, the Joint FAO/WHO Expert Committee on Food Additives (JECFA, Rome, Italy) and China established maximum residue limits of CIP in animal-derived food to prevent the accumulation of antimicrobial residues, e.g., 100 μg/kg in milk and meat.

By now, many physicochemical methods have been reported for the detection of residues of FQs in foods of animal origin. These analytical methods are highly sensitive and dependable; however, such methods require specialized instrumentation, trained personals, and are time consuming. They are unsuitable for the rapid evaluation of large numbers of samples. Immunoassays, especially the indirect competitive enzyme-linked immunosorbent assay (IC-ELISA), which is based on the principle that antibodies specifically bind to antigens, are considered the most reliable method for detecting antibodies [6,7]. In previous studies, researchers have developed IC-ELISA based on monoclonal antibodies (MAbs) to determine fluoroquinolone in food of animal origin [8–10]. Although ELISA is a mature and widely used method, it has many rigorous programs for preparing traditional antibodies (PAbs and MAbs) from antigen-immunized animals [11]. Hence, a simple, rapid, and effective technology for preparing novel antibodies must be developed.

The development of gene engineering techniques facilitated the production of various gene recombinant antibodies, and single-chain variable fragment (scFv) is the most popular format of recombinant antibody that has been successfully constructed by assembling the variable-heavy (VH) region and light chain (VL) domain of an antibody with a flexible linker [12]. The intrinsic properties of scFv antibodies can be improved by various mutagenesis techniques [13]. The recognition property of an scFv antibody can be evolved in vitro [14]. For the evolution of the scFv antibody, its recognition mechanism should be studied first, and binding sites, contact amino acids, and intermolecular forces should be determined [15]. In recent years, molecular docking has been used in analyzing the interactions between ligands and scFv antibodies, and random mutagenesis and site-directed mutagenesis have been used in obtaining scFv mutants [16,17].

Phage display technology (PDT) is the integration of foreign genes into specific coat protein genes of phage and fusion, with coat protein to promote ligand recognition and binding [18,19]. It is considered to be the most suitable technology for the production of single-chain antibodies. The phage antibody library uses genetic engineering methods to amplify VH and VL genes. After random combination, it is inserted into the phage coat protein gene and fused and expressed on the surface of the phage [20]. Specific single-chain antibodies are obtained through specific panning, which is extensively used for preparing antigen-specific artificial antibodies in biomedicine, environmental pollutants analysis, and food safety detection fields. For example, Xu et al. [21] and Zhao et al. [22] obtained the broad-specificity domain antibodies for Bt Cry toxins and pyrethroid pesticides by rounds of specific phage library biopanning, respectively, which are all based on phage antibody library technology.

In this study, an immunized mouse phage display scFv library for screening of anti-CIP phage scFv particles was constructed. Then, we transfected the phage to *E. coli* BL21 for expression, to obtain a highly sensitive anti-CIP scFv. The scFv recognition mechanism was studied through molecular docking, and the sensitivity and cross-reactivity were improved through targeted mutagenesis. Then, IC-ELISA was developed based on the scFv mutant to detect the CIP in animal-derived edible tissues.

2. Material and Methods

2.1. Reagents and Chemicals

Ciprofloxacin (CIP), enrofloxacin (ENR), sarafloxacin (SAR), difloxacin (DIF), lomefloxacin (LOM), enrofloxacin (ENO), norfloxacin (NOR), amifloxacin (AMI), marbofloxacin (MAR), danofloxacin (DAN), fleroxacin (FLE), ofloxacin (OFL) and pefloxacin (PEF) were obtained from the China Institute of Veterinary Drug Control (Beijing, China). All chemicals and reagents used in this study were at least analytical grade or better. The standard stock solutions of these FQs were prepared with methanol (10 μg/mL), and their working solutions with series concentrations (0.1–200 ng/mL) were diluted from the stock solutions with PBS. All the standard solutions were stored at 4 °C to remain stable for 8 weeks. N-hydroxy succinimide (NHS), γ-aminobutyric acid (4AS), bovine serum albumin (BSA), ovalbumin (OVA), 1-ethyl-3- (3-dimethylaminopropyl)-carbodiimide (EDC), goat anti-mouse IgG horseradish peroxidase conjugate (HRP-IgG), Freund's complete adjuvant (FCA) and Freund's incomplete adjuvant (FIA) were from Sigma (St. Louis, MO, USA). PBS (pH 7.2) was prepared by dissolving 0.2 g of KH_2PO_4, 0.2 g of KCl, 1.15 g of Na_2HPO_4, and 8.0 g of NaCl in 1000 mL of deionized water. Washing buffer (PBST) was PBS buffer containing 0.05% Tween. Coating buffer was 5% MPBS (5% Skim milk powder in PBS). Substrate buffer was 0.1 mol/L citrate (pH 5.5). The substrate system was prepared by adding 200 μL of 1% (*w*/*v*) TMB in DMSO and 64 μL of 0.75% (*w*/*v*) H_2O_2 into 20 mL of substrate buffer.

All the restriction enzymes and DNA modification enzymes were molecular biology grade. The RNase prep pure Cell/Bacteria Kit was from Tiangen Biotech Co. Ltd. (Beijing, China). The Prime script RT-PCR Kit, IPTG (isopropyi-β-D-thirgalactopyranoside), X-Gal, pCANTAB5E Vector Cloning kit, horseradish peroxidase-labeled goat anti-GST-tag antibody, restriction enzymes (Sfi I and Not I) and T4 DNA Ligase were from Takara Company (Dalian, China). The EasyPure Quick Gel Extraction Kit, Easy Pure Plasmid Miniprep Kit, express vector PET-32a competent cell BL21(DE3), Fast MultiSite Mutagenesis System and Luria–Bertani culture medium (liquid and solid) were from TransGen Biotech (Beijing, China). The DNA Purification Kit and SDS-PAGE gel preparation kit were from Beijing ComWin Biotech Co. Ltd. (Beijing, China). The synthesis of primers and the analysis of gene sequence were performed at Sangon Biotechnology Co. Ltd. (Shanghai, China).

2.2. Synthesis of Antigen

The immunogens CIP-BSA and coating antigens CIP-OVA were synthesized in this study. The details are described below. A mixture of CIP (30 mg), NHS (25 mg) and EDC (30 mg) in 1.5 mL of N, N-dimethylformamide (DMF) was stirred at room temperature overnight. Then, the activated CIP was centrifuged for 15 min (5000 rpm), and the supernatant was added dropwise to 70 mg of BSA dissolved in a solution consisting of 10 mL of PBS and 1 mL of DMF under stirring. The conjugation mixture was stirred at 4 °C for 5 h, and then centrifuged for 10 min (5000 rpm). The supernatant was dialyzed against 0.01 mol/L PBS for 72 h. The dialysis solution was stored at −20 °C. The coating antigens CIP-OVA were prepared as described in the CIP-BSA synthesis section, except that BSA was replaced by OVA.

2.3. Immunization

All animal experiments in this study adhered to the Zhengzhou University animal experiment center guidelines and were approved by the Animal Ethics Committee. Six Balb/c female mice (8 weeks old) were induced to express anti-CIP MAbs by immunizing the mice with five rounds of subcutaneous injection of CIP-BSA conjugates. In the first round of immunization, 250 μg of CIP-BSA with FCA was emulsified for subcutaneous multipoint injection, then four subsequent injections were given at 14-day intervals that were emulsified in FCA. Antisera were collected 7 days after the third and fourth immunization, and the antibody titer was determined through indirect ELISA. A week after the fourth round of immunization, booster immunization with 150 μg of CIP-BSA was performed. After 5 days, blood and spleen samples were collected for the construction of the phage display scFv library.

2.4. Phage Display scFv Library Construction

Total RNA was extracted from mouse tissues with TRIzol reagent according to the manufacturer's instructions. Then, total RNA was used as a template in the reverse transcription of cDNA. The sequences of the primers were used in the amplification of the cDNAs of VH and VL genes for scFv construction. The primers used for the amplification of scFv coding sequences were designed according to Table 1 and then spliced to a whole scFv gene through splicing overlap extension PCR (SOE-PCR). The system conditions were as follows: 94 °C for 5 min, 30 cycles at 94 °C for 45 s, 58 °C for 60 s, and 72 °C for 45 s, and final extension at 72 °C for 10 min. Gene fragments encoding VH and VL were amplified and spliced to a single gene by using a DNA linker encoding a pentadeca peptide (Gly4Ser) 3 through primary PCR. The system conditions were 94 °C for 5 min, 30 cycles of at 94 °C for 45 s, 60 °C for 60 s, 72 °C for 45 s, and final extension at 72 °C for 10 min. PCR products were verified through agarose gel electrophoresis, and the relevant fragments were sequenced. The gene fragments were then digested with Sfi I and Not I restriction endonuclease and ligated into pCANTAB5E phagemid vectors. The recombinant vectors were then transformed into *E. coli* TG1 cells. Serial dilutions of 10−1−10−8 were plated onto SOB plates (2% tryptone, 0.5% yeast extract, 0.05% NaCl, 2.5 mM KCl, 10 mM $MgCl_2$, and 1.5% Agar powder) that contained 100 μg/mL Amp and 2% Glu. After inoculation, all the plates were incubated overnight in a previously set incubator at 30 °C, then clones were randomly selected and screened for inserts by performing another round of PCR. Finally, the colonies were scraped into 20 mL of 2YT (1.6% Tryptone, 1% yeast Extract, and 0.5% NaCl), named the original antibody library, and stored at −80 °C in 20% glycerol.

Table 1. Nucleotide primer sequences.

Primer Names	Nucleotide Sequences (5′→3′)
VH for	GCGGCCCAGCCGGCCATGGCCGARGTGAAGCTGGTGGARTCTGGR
VH back	AGCGGCGGTGGCGGTTCTGGAGGCGGCGGTTCTGAYATGCAGATGACMCAG
VL for	AGCGGCGGTGGCGGTTCTGGAGGCGGCGGTTCTRAMATTGTGMTGACCCAATC AGCGGCGGTGGCGGTTCTGGAGGCGGCGGTTCTGAYATGCAGATGACMCAGWC
VL back	ACTAGTCGCGGCCGCGTCGACAGCMCGTTTBAKYTCTATCTTTGT ACTAGTCGCGGCCGCGTCGACAGCMCGTTTCAGYTCCARYTT
scFv for	CGCAATTCCTTTAGTTGTTCCTTTCTATGCGGCCCAGCCGGCCATGGCC
scFv back	GGTTCCAGCGGATCCGGATACGGCACCGGACTAGTCGCGGCCGCGTCGAC

2.5. *Phage scFv Particle Enrichment and Screening*

The phage library underwent four rounds of biopanning with coat antigen CIP-BSA for phage scFv particle enrichment. A sterile cell flask was coated with 2 mL of CIP-OVA (the first round was 50 μg/mL, and the remaining three rounds were 25, 12, and 6 μg/mL) in PBS solution at 4 °C and left to stand overnight. The flask was washed five times with PBST solution and blocked with MPBS at 37 °C for 2 h. After being washed with PBST solution, 1 mL of library phage particles was added into a flask for shaking for 1 h at 150 rpm at room temperature, then left to stand for 1 h. The CIP-OVA-bound phage scFv particles were washed with PBST solution and eluted with 1 mL of trypsin solution (1 mg/mL in PBS). The eluent was the first round of enrichment library, and the phage scFv particles were amplified for the next round of enrichment. Four rounds of biopanning were performed. The fourth round of enriched anti-CIP phage particles was infected with *E. coli* TG1 and spread on a TYE-AG medium (contains 100 μg/mL Amp and 1% Glu) for culturing overnight at 37 °C. Individual colonies were randomly picked and grown in 2 × TY-AG medium glucose with 100 μg/mL ampicillin for 16 h at 37 °C and 200 rpm. The next day, 10 μL of culture per well was transferred into another 96-well plate for culturing for 2 h at 37 °C and 200 rpm, and M13KO7 helper phages were added to rescue for 2 h at 37 °C and 200 rpm. The plate was centrifuged at 3300 rpm for 20 min at 37 °C, and the pellets were resuspended with 250 μL/well of 2 × TY-AK medium and cultured overnight at 30 °C and 200 rpm. Finally, the plate was centrifuged at 4 °C and 3300 rpm for 30 min, then the supernatant was used in the monoclonal phage ELISA for CIP.

2.6. *Colony PCR and Sequencing*

The positive phage scFv colonies were cultured in a 2 × TY-AG medium until the logarithmic phase for colony PCR, and the PCR products were examined by 1% agarose gel electrophoresis. The selected positive monoclonal phages were sequenced by Sangon Biotechnology (Shanghai, China) Co., Ltd.

2.7. *Expression and Purification of scFv*

The target gene and prokaryotic expression vector pET-32a were digested with NcolI and NotI restriction enzymes and linked using T4 DNA ligase. Then, the positive recombinant plasmid was used in producing *E. coli* strain BL21 (DE3). The mixture was heat shocked for 90 s at 42 °C, and cultured in a Luria–Bertani (LB) medium (1% tryptone, 0.5% yeast extract, and 1% NaCl) containing 100 μg/mL kanamycin at 37 °C overnight. After the OD_{600} of the bacterium solution reached 0.6–0.8, 1 mmol/L IPTG was added to the culture to induce the expression of scFv. The culture was further grown at 37 °C for 16 h. The supernatant was collected and concentrated 100-fold by using MWCO: 8000–14,000 Da of dialysis bag in PEG/NaCl. The collected pellets were resuspended with PBS for the production of a periplasmic lysate and lysed through sonication for the production of the whole-cell lysate. The supernatant and periplasmic and whole-cell lysates were used in analyzing the solubility of the proteins through SDS-PAGE. Finally, BioMag-SA GST-tag Protein Purification magnetic beads were used to purify the anti-CIP scFv protein.

2.8. Denaturation and Renaturation of the scFv Protein

The inclusion bodies were washed five times with PBS containing 0.1% TritonX-100 and 2 mol/mL urea at 2 h intervals; then, the inclusion bodies were solubilized in 20 mL of 8 mol/mL urea solutions and slowly stirred at 4 °C for 16 h. The solubilized solution was centrifuged for 20 min (12,000 rpm). Finally, the solution containing denatured scFv was dialyzed in PBS at 4 °C for 48 h for the removal of urea from the protein solutions.

2.9. Characterization of scFv Antibody

Western blot. A volume of scFv solution was added to a nitrocellulose membrane immersed in blocking buffer (4% BSA in PBS (*w*/*v*)) for 1 h. Then, a volume of horseradish peroxidase-labeled anti-GST-tag antibody (1:2000) was added to the block point, and the membrane was incubated for 2 h at room temperature. Finally, a volume of substrate solution (4-chloro-1-naphthol) was added for the visualization of the result.

Indirect competitive ELISA. The purified anti-CIP scFv was used in establishing IC-ELISA. Briefly, 100 μL/well of CIP-OVA solution was coated into 96-well plates overnight at 4 °C; then, the plates were washed with PBST solution and blocked with 300 μL per well of 5% MPBS at 37 °C for 1 h. scFv (100 μL/well) previously diluted with PBS and a series of CIP standard concentrations (200, 100, 80, 50, 20, 10, 5, 2, 1, and 0.1 ng/mL) were washed with PBST and then mixed. The plates were incubated at 37 °C for 1 h, then washed with PBST. Avidin conjugated with horseradish peroxidase (100 μL/well; 1/2000 dilution in PBS) was added to the wells and incubated at 37 °C for 30 min. The wells were then washed five times with PBST, and 100 μL/well of TMB substrate was added and incubated for 10 min in the dark at room temperature. Color reaction was stopped with the addition of sulfuric acid (2 mol/L, 50 μL/well). Finally, absorbance was measured at 450 nm with an automatic microplate reader (Thermo, Waltham, MA, USA). The IC50 value, assay dynamic range, and limit of detection (LOD) served as the criteria for evaluating IC-ELISA. The inhibition ratios of anti-CIP scFv, IC10, IC20, IC50, and IC80 were calculated using the formula [(P-S-N)]/(P-N)] × 100%, where P is the OD_{450} value of the positive sample (50 μL of anti-CIP scFv mixed with 50 μL of CBS), S is the OD_{450} value of the standard (50 μL of scFv mixed with 50 μL of the serial concentration of CIP), and N is the OD_{450} value of the negative control (100 μL of CBS).

2.10. Homology Modeling and Molecular Docking

In this experiment, the possible template sequences of the anti-CIP scFv model were searched in the NCBI database (https://www.ncbi.nlm.nih.gov/ (accessed on 1 February 2021)), and sequence comparison was performed in the BLAST section for a selection template of a high consistency with the anti-CIP scFv model. The anti-CIP scFv template sequence is as follows: (sense): 5′-TCAAGTGTAAGTTACATGCCATGGTACCAGCAG-3′ and (antisense): 5′-TCTTGGCTTCTGCTGGTACCATGGCATGTAACTTACACT-3′. The sequence with a high score and low e-value was used as a template sequence for anti-CIP scFv model building. Then, the SWISS MODEL online server was used in the homology modeling of anti-CIP scFv. To verify the reliability of the homology modeling results and determine the best model structure, we used Procheck, Verify3D, and ERRAT programs in the evaluation of the consistency of the constructed an anti-CIP scFv homology model and selected the best receptor model for further molecular docking study. To study the binding mode of CIP with anti-CIP-scFv and find key residues, we used MOE 2015.10 in exploring the molecular docking of CIP with scFv. In the Dock module, CIP was docked into the active site of anti-CIP-scFv through the method of Induced Fit and under Amber10: EHT forcefield. The docking ligand, which had 30 docking conformations after default parameters were used, were used for further analysis.

2.11. *Directional Mutagenesis of scFv Antibody*

The binding affinity of CIP with scFv-CIP antibody was improved through the virtual mutation of the potential key residues of scFv-CIP. The process was based on the study of the binding mode of anti CIP scFv with CIP. MOE 2015.10 software was used in conducting the virtual mutation of amino acid residues that affect the binding of CIP with scFv and directly replace them with other amino acids. In this study, Ser was used to replace Val160 for the production of the mutant of scFv-CIP antibody. Then, the structure of the virtual mutation scFv model was optimized, and a stable scFv mutation model was obtained. Subsequently, the docking study of CIP with scFv mutation was performed through the method of Induced Fit, and 30 conformations were obtained using the default parameters. During the experiments, the scFv gene in the express vector scFv-pCANTAB5E was mutated directly for the production of a mutated express vector with a fast-multisite mutagenesis system according to the manufacturer's recommended protocol. Then, the mutated express vector was expressed for the production of scFv mutant using the procedures described above. The scFv mutant was identified and analyzed through SDS-Page and IC-ELISA [6].

2.12. *Sample Preparation and Cross-Reactivity Analysis*

Beef, pork, milk, and chicken samples were obtained from a local market. CIP (1000 μg/mL, prepared in PBS) was added to each sample for the production of spiked concentrations of 0, 50, 100, and 200 μg/kg. Aliquots of the homogenized tissue samples (1 g of wet mass) were transferred to a 50 mL centrifuge tube. Exactly 5 mL of 5% trichloroacetic acid and 10 mL of 0.2 M PBS were mixed with the tissue sample, and the mixture was incubated for 1 h at 60 °C. Subsequently, the suspension was centrifuged at 5000× *g* for 20 min. The supernatant was separated and diluted tenfold with deionized water. The aliquots (100 mL each) were distributed into the microtiter plate. The CIP standards of different concentrations (0, 50, 100, and 200 μg/L) were added to milk samples, which were then defatted by centrifugation at 5000× *g* for 20 min at 4 °C. After 60 μL of sodium nitroprusside (0.36 mol/L) and 60 μL of zinc sulfate (1.04 mol/L) were added to 2 mL of each defatted milk sample, the samples were vortexed for 1 min and then centrifuged at 5000× *g* for 20 min at 4 °C. The supernatant was removed and diluted tenfold with PBS for analysis. Recoveries were calculated on the basis of the standard curve constructed by IC-ELISA.

The specificity of the scFv under optimized conditions was evaluated by measuring cross-reactivity (CR) with a group of structurally related compounds, including 12 other analogs such as enrofloxacin, danofloxacin, and fleroxacin. The CRs of anti-CIP scFv for CIP analogues were calculated using the formula: [CR (%) = IC50 (CIP)/IC50 (CIP analogue)] × 100%.

2.13. *Statistical Analysis*

The statistical software SPSS 16.0 and data processing system 14.0 (DPS) were used for statistics. Values are expressed as mean ± standard deviation. All data are suitable for analysis without any conversion.

3. Results and Discussion

3.1. Construction of Phage Display scFv Library

Compared with conventional antibodies (monoclonal antibodies and polyclonal antibodies), scFv can be produced on a large scale in prokaryotic and eukaryotic systems, so it is cheap and saves time [23,24]. In addition, the scFv antibody can be studied at the molecular level (homologous modeling and molecular docking), and its antigen binding affinity can be improved through gene mutation and gene reforming [15]. In this study, a scFv library for mouse phage display was constructed, and total RNA was extracted from the spleen of immunized mice and then reverse-translated to cDNA. The VH and VL coding sequences were amplified using the cDNA as the template, and a complementary linker sequence was added. The amplified and purified NcoI-VL-linker and linker-VH-NotI were spliced to whole scFv genes through SOE-PCR. As shown in Figure 1, the amplified VH, VL, and scFv DNAs were approximately 350, 330, and 780 bp long, respectively. The purified scFv and pCANTAB5E vectors were digested with SfiI and NotI. T4 DNA ligase was used to ligate the products, and recombinant plasmids were translated into *E. coli* TG1 cells, and a library with a capacity of 3.34×10^9 CFU/mL was constructed successfully.

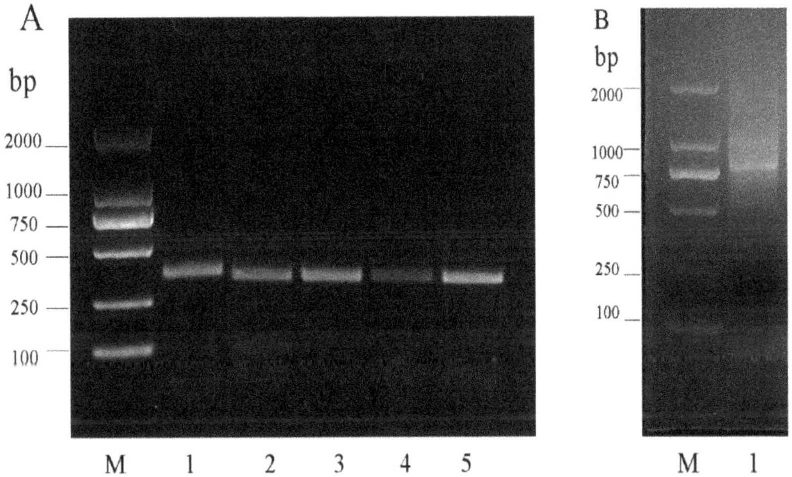

Figure 1. Amplification of heavy and light chains. Agarose gel electrophoresis of the amplified antibody variable fragments. (**A**) PCR amplification of VH and VL. Lane 1: PCR products of VH, about 350 bp. Lane 2. Lane 3 and Lane 4: PCR products of VL, about 330 bp. (**B**) Amplification of scFv by overlap PCR, about 780 bp.

3.2. Panning of Phage-Displayed Antibody Libraries

For the production of highly specific antibodies, the washing steps were progressively increased, whereas the concentrations of coated CIP-OVA were decreased (Table 2) as described by Li et al. [6]. It can be seen from Figure 2A that after the first three rounds of panning, the antibody response signal gradually increased, and decreased after the fourth round of panning, indicating that the phage antibody library was effectively enriched with specific phage particles after four rounds of panning. On the plate after the fourth round of panning, 25 phage colonies were randomly selected for Phage-ELISA to analyze the binding ability of ciprofloxacin. The results are shown in Figure 2B. Clone scFv-22, which showed relatively stable and high binding abilities, was selected for further study.

Table 2. The library size and phage titer of each panning.

Rounds	Coated Antigen	Coating Concentrations (μG/WELL)	Input	Output	Output/Input
1	CIP-OVA	50	4×10^{11}	2.9×10^{5}	7.55×10^{-7}
2	CIP-OVA	25	4×10^{11}	5.6×10^{6}	1.65×10^{-5}
3	CIP-OVA	12	4×10^{11}	2.9×10^{8}	5.9×10^{-4}
4	CIP-OVA	6	4×10^{11}	3.34×10^{9}	6.15×10^{-3}

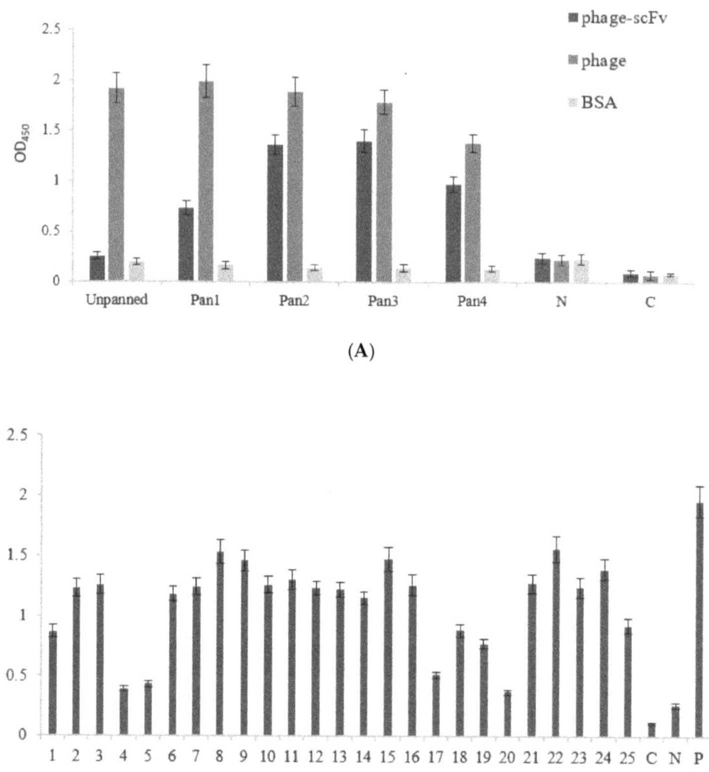

Figure 2. Phage-ELISA. (**A**) The enrichment of specific scFv in each library after four rounds panning. (**B**) Binding activity of scFv antibodies to CIP. 1–25: scFv antibodies from randomly selected clones from the 4th panning; C: blank control; N: negative control, BSA; P: positive control, cell supernatant.

3.3. Expression, Purification of the scFv-22

As is well known, IPTG concentration, post-induction time, and incubation temperature are the main factors for optimizing protein expression [25]. In the preliminary experiments of this study, the highest scFv expression level was obtained at the following conditions: 37 °C, 16 h, and 1 mM IPTG. Under optimal expression conditions, the scFv-22 antibody fragments were expressed in *E. coli* HB2151, and proteins were determined through Western blotting, as shown in Figure 3. The results indicated that the scFv-22 antibody (approximately 47 kDa) was expressed successfully. Anti-CIP scFv proteins were purified using BioMag-SA GST-tag Protein Purification magnetic beads according to the manufacturer's instructions. The purity of protein solution was confirmed through SDS-PAGE, as shown in Figure 4, and then the purified protein was stored at −2 °C.

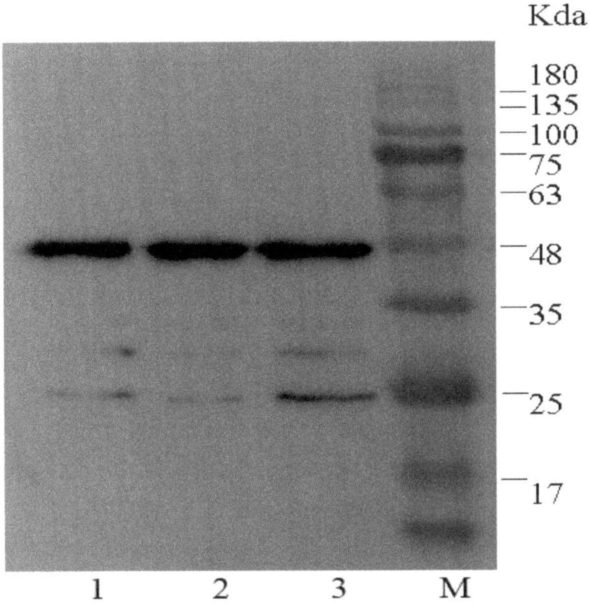

Figure 3. Analysis of scFv-22 by Western blot. Lane 1–3: scFv-22 induced expression. Lane M: Protein 180 marker.

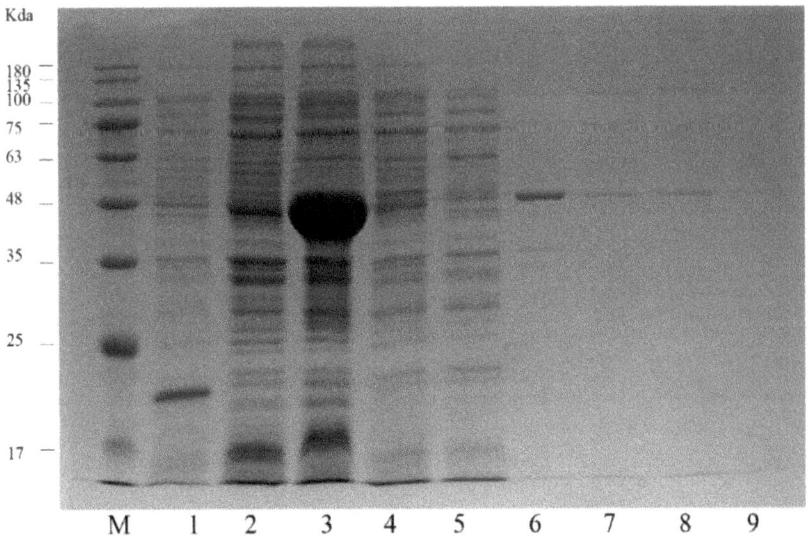

Figure 4. Analysis of scFv by SDS-PAGE. Lane M: Protein 180 marker. Lane 1: PET-32a vector. Lane 2–3: After induction of expression, the supernatant and precipitate obtained by sonication. Lane 4–5: The supernatant obtained after combining with the magnetic beads. Lane 6–9: Supernatant after washing with 20 mM, 100 mM, 200 mM, 250 mM imidazole, respectively.

3.4. IC-ELISA for CIP and Its Analogues Based on scFv-22

The performance of the purified scFv-22 was evaluated using IC-ELISA. The optimum concentration of scFv-22 was 0.25 μg/mL, producing an OD_{450} of 1.0 at 4 μg/mL of CIP-

OVA coating concentration through checkerboard titration. Under the optimal conditions, the regression curve equation of CIP-scFv was $y = -0.4517x + 1.1409$ ($R^2 = 0.9877$), as shown in Figure 5. The IC50 value of the assay established with scFv-22 was 26.23 ng/mL, demonstrating that scFv can be used in detecting CIP. The linear range of the assay established with scFv ranged from 5.68 ng/mL to 201.55 ng/mL.

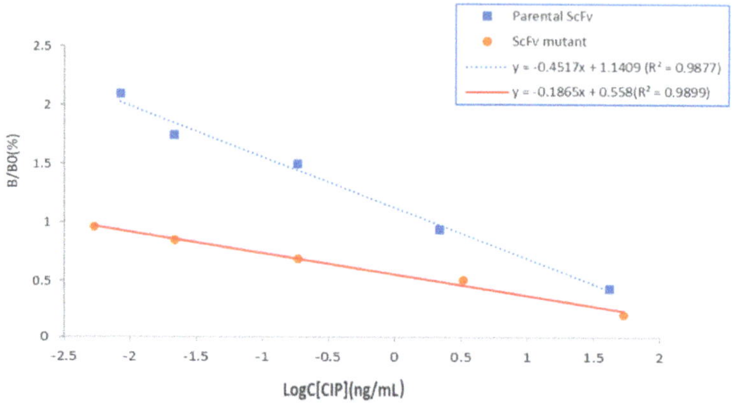

Figure 5. Standard curves of the competitive ELISA for CIP. X-axis shows the logarithm concentration of CIP, Y-axis represents the inhibition rate (B/B0). B0 and B are the absorbance values obtained from binding at zero and certain concentrations of CIP standard.

3.5. Homology Modeling and Molecular Docking

In this experiment, given the results of parameter evaluation, such as sequence identity and structural similarity, the protein PDB ID: 3UZQ was selected as the template sequence, which has the highest sequence identity (78.8%) with the target sequence anti-CIP-scFv. Subsequently, a stable model of single-chain antibody was established using the SWISS MODEL homology modeling software. The single-chain antibody of anti-CIP-scFv was connected by VH and VL through three connecting peptides (Gly4Ser) and had a typical single-chain antibody structure with anti-parallel β-sheet and loop regions.

We evaluated the model to verify its reliability. The Procheck program was used in evaluating the three-dimensional structure of the anti-CIP-scFv model. The Ramachandran plot showed that 91.7% of the amino acids residues in the model were located in the core region, 7.8% in the allowable region, and only 0.5% in the forbidden zone of the twist angle. The results showed that the dihedral angles of 99% of protein residues in scFv model were within the reasonable range and conformed to the rule of stereochemical energy. The result of Verify-3D showed that the average 3D–1D score of 99.58% of amino acid residues in the scFv model was greater than 0.2, and the model passed the Verify-3D test. The Errat result showed that the overall quality factor was 87.82. Therefore, the experimental model of anti-CIP-scFv has high reliability and can be used in the molecular docking of the CIP antigen.

The docking result of scFv-22 with CIP is shown in Figure 6. The active site of scFv-22 consists of residues Gln153, Pro155, Ala156, Leu158, Val160, Ile168, Val229, Asp233, Ala234, Ala235, Thr236, and Tyr237. The carboxyl group of CIP can form a 2.78 Å and 3.15 Å hydrogen bond with the residue Gly251. The quinoline structure is located in the hydrophobic cavity formed by residues Gln153, Pro155, Ala156, Ile168, Gly251, and Thr236. The cyclopropyl group can interact with the residues Ala235 and forms hydrophobic interaction with Tyr238. Thus, these forces may be the main reason for the increased ability to bind to CIP. However, the hydrophilic carboxyl group in CIP and the carbonyl group on the quinoline are close to the hydrophobic residue Val160. We preliminarily speculated that if the residue Val160 is transformed to a hydrophilic amino acid, it will promote the combination of CIP to scFv. When Val160 was substituted by Ser, the total binding energy decreased from −5.23 to −7.91 kcal/mol, and the number of hydrogen bonds and the amino acids forming hydrophobic interaction all increased in the binding site (Figure 6B,D), indicating that the intermolecular forces of scFv-CIP increased. Therefore, Val160 was substituted with Ser for the directional mutagenesis of the scFv antibody in the present study.

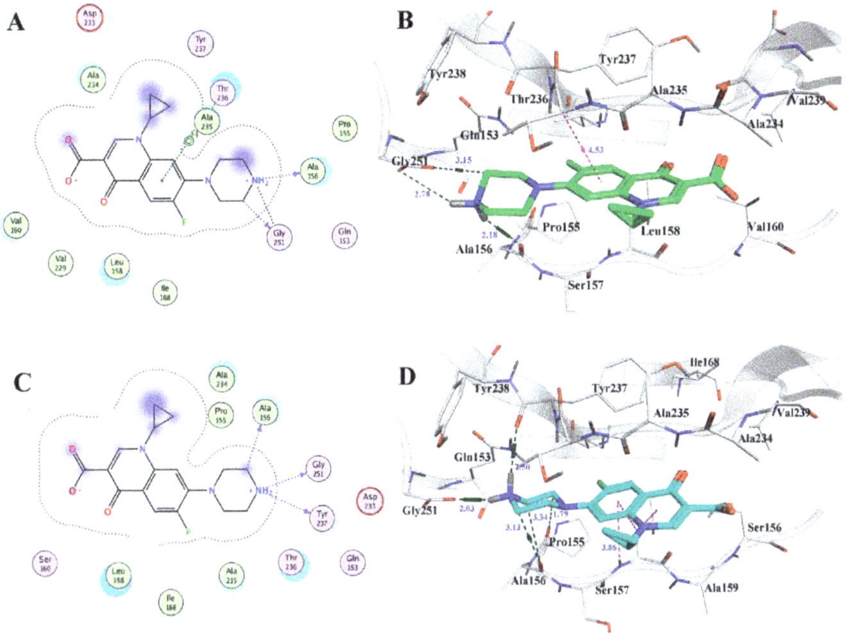

Figure 6. (**A**) The binding mode between parental scFv and CIP, (**B**) the interaction between parental scFv and CIP binding site; (**C**) the binding mode between scFv mutant and CIP, (**D**) the interaction between scFv mutant and CIP binding site.

3.6. Characterization of Mutant scFv

The performance of purified scFv mutant was evaluated through SDS-PAGE and IC-ELISA. Under the same conditions as the parental scFv, the regression curve equation of the mutant scFv was y = −0.1865x + 0.558 (R^2 = 0.9899), as shown in Figure 5. The IC50 value of the assay established with the scFv mutant was 1.58 ng/mL, indicating that the affinity of the scFv mutant increased 16.6 times compared with that of the parental scFv. The mutant had a higher affinity and better sensitivity than the original antibody, indicating that the parental scFv antibody was evolved successfully.

3.7. Precision and Recovery

Spiking and recovery tests were conducted for the assessment of the feasibility of IC-ELISA. During the tests, the mutant scFv was used in detecting CIP in spiked samples, and no positive results were obtained for the non-spiked samples. All samples spiked with CIP showed good agreement between the spiking level and concentration detected, as shown in Figure 3. In the intra-assay, the mean recovery for CIP ranged from 73.80% to 121.58% and the RSD values ranged between 2.01% and 7.35% (based on triple measurements within a day). In the inter-assay, the mean recovery for CIP ranged from 75.29% to 123.35%, and RSD values ranged between 1.49% and 9.81% (based on triple measurements in 3 days). As demonstrated with the samples spiked with CIP, the IC-ELISA method provided satisfactory results for the detection of CIP residues in milk and food animal tissues. The cross-reactivity of the CIP-scFv with danofloxacin, enrofloxacin, and fleroxacin and many more were tested through IC-ELISA. As shown in Figure 4, the CIP-scFv showed low cross-reactivity with other fluoroquinolones, indicating that scFv is highly specific for CIP.

4. Conclusions

A highly sensitive anti-CIP single-chain antibody was obtained through phage display and directional evolution, and a rapid and highly sensitive IC-ELISA method for detecting CIP residues in products of animal origin was developed. The method showed good stability, reproducibility, and accuracy for detecting CIP, indicating a wide application prospect for the rapid and sensitive detection of antibiotic residues in animal-derived food.

Author Contributions: Conceptualization, F.W. and N.L.; methodology, N.L.; software, F.W.; validation, Y.Z. (Yunshang Zhang) and M.H.; formal analysis, Y.Z. (Yunshang Zhang); investigation, X.S.; resources, X.S.; data curation, F.W.; writing—original draft preparation, Y.Z. (Yunshang Zhang); writing—review and editing, F.W.; visualization, Y.Z. (Yali Zhao); supervision, N.L. and J.F.; project administration, F.W.; funding acquisition, F.W. All authors have read and agreed to the published version of the manuscript.

Funding: This research was funded by the National Key Research and Development Program of China (No. 2018YFC1602902, 2019YFC1605701).

Informed Consent Statement: Not applicable.

Acknowledgments: The authors are grateful for the financial supports from the National Key Research and Development Program of China (No. 2018YFC1602902, 2019YFC1605701).

Conflicts of Interest: The authors declare no conflict of interest.

References

1. Dalhoff, A. Antiviral, antifungal, and antiparasitic activities of fluoroquinolones optimized for treatment of bacterial infections: A puzzling paradox or a logical consequence of their mode of action? *Eur. J. Clin. Microbiol. Infect. Dis.* **2015**, *34*, 661–668. [CrossRef] [PubMed]
2. Bird, S.T.; Etminan, M. Risk of acute kidney injury associated with the use of fluoroquinolones. *CMAJ* **2013**, *185*, E475–E482. [CrossRef] [PubMed]
3. Patel, K.; Goldman, J.L. Safety Concerns Surrounding Quinolone Use in Children. *J. Clin. Pharmacol.* **2016**, *56*, 1060–1075. [CrossRef]
4. Li, J.; Hao, H. The effects of different enrofloxacin dosages on clinical efficacy and resistance development in chickens experimentally infected with Salmonella Typhimurium. *Sci. Rep.* **2017**, *7*, 11676. [CrossRef]
5. Xu, L.; Wang, H. Integrated pharmacokinetics/pharmacodynamics parameters-based dosing guidelines of enrofloxacin in grass carp Ctenopharyngodon idella to minimize selection of drug resistance. *BMC Vet. Res.* **2013**, *9*, 126. [CrossRef] [PubMed]
6. Cui, L.; Jinxin, H. Preparation of a Chicken scFv to Analyze Gentamicin Residue in Animal Derived Food Products. *Anal. Chem.* **2016**, *88*, 4092–4098.
7. Abdelwahab, M.; Loa, C.C. Recombinant nucleocapsid protein-based enzyme-linked immunosorbent assay for detection of antibody to turkey coronavirus. *J. Virol. Methods* **2015**, *217*, 36–41. [CrossRef] [PubMed]
8. Huang, B.; Yin, Y. Preparation of high-affinity rabbit monoclonal antibodies for ciprofloxacin and development of an indirect competitive ELISA for residues in milk. *J. Zhejiang Univ. Sci. B* **2010**, *11*, 812–818. [CrossRef]
9. Fan, G.-Y.; Yang, R.-S. Development of a class-specific polyclonal antibody-based indirect competitive ELISA for detecting fluoroquinolone residues in milk. *J. Zhejiang Univ. Sci. B* **2012**, *13*, 545–554. [CrossRef]
10. Zhang, H.-T.; Jiang, J.-Q. Development of an indirect competitive ELISA for simultaneous detection of enrofloxacin and ciprofloxacin. *J. Zhejiang Univ. Sci. B* **2011**, *12*, 884–891. [CrossRef] [PubMed]
11. Li, C.; Luo, X. A Class-Selective Immunoassay for Sulfonamides Residue Detection in Milk Using a Superior Polyclonal Antibody with Broad Specificity and Highly Uniform Affinity. *Molecules* **2019**, *24*, 443. [CrossRef]
12. Makvandi-Nejad, S.; Sheedy, C. Selection of single chain variable fragment (scFv) antibodies from a hyperimmunized phage display library for the detection of the antibiotic monensin. *J. Immunol. Methods* **2010**, *360*, 103–118. [CrossRef] [PubMed]
13. Norihiro, K. Anti-estradiol-17beta single-chain Fv fragments: Generation, characterization, gene randomization, and optimized phage display. *Steroids* **2008**, *73*, 1485–1499.
14. Kobayashi, N.; Oyama, H.; Kato, Y.; Goto, J.; Söderlind, E.; Borrebaeck, C.A. Two-step in vitro antibody affinity maturation enables estradiol-17beta assays with more than 10-fold higher sensitivity. *Anal. Chem.* **2010**, *82*, 1027–1038. [CrossRef]
15. Liu, J.; Zhang, H.C. Production of anti-amoxicillin ScFv antibody and simulation studying its molecular recognition mechanism for penicillins. *J. Environ. Sci. Health Part B Pestic. Food Contam. Agric. Wastes* **2016**, *51*, 742–750. [CrossRef] [PubMed]

16. Wen, K.; Nolke, G. Improved fluoroquinolone detection in ELISA through engineering of a broad-specific single-chain variable fragment binding simultaneously to 20 fluoroquinolones. *Anal. Bioanal. Chem.* **2012**, *403*, 2771–2783. [CrossRef] [PubMed]
17. Tao, X.; Chen, M. Chemiluminescence competitive indirect enzyme immunoassay for 20 fluoroquinolone residues in fish and shrimp based on a single-chain variable fragment. *Anal. Bioanal. Chem.* **2013**, *405*, 7477–7484. [CrossRef]
18. Kumar, R.; Parray, H.A. Phage display antibody libraries: A robust approach for generation of recombinant human monoclonal antibodies. *Int. J. Biol. Macromol.* **2019**, *135*, 907–918. [CrossRef]
19. Zhao, A.; Tohidkia, M.R.; Siegel, D.L.; Coukos, G.; Omidi, Y. Phage antibody display libraries: A powerful antibody discovery platform for immunotherapy. *Crit. Rev. Biotechnol.* **2016**, *36*, 276–289. [CrossRef]
20. Arap, M.A. Phage display technology: Applications and innovations. *Genet. Mol. Biol.* **2005**, *28*, 1–9. [CrossRef]
21. Xu, C.; Miao, W. Construction of an immunized rabbit phage display antibody library for screening microcystin-LR high sensitive single-chain antibody. *Int. J. Biol. Macromol.* **2019**, *123*, 369–378. [CrossRef]
22. Zhao, Y.; Liang, Y. Isolation of broad-specificity domain antibody from phage library for development of pyrethroid immunoassay. *Anal. Biochem.* **2016**, *502*, 1–7. [CrossRef]
23. Zhang, X.; Zhang, C. Construction of scFv phage display library with hapten-specific repertories and characterization of anti-ivermectin fragment isolated from the library. *Eur. Food Res. Technol.* **2010**, *231*, 423–430. [CrossRef]
24. Chaisri, U.; Chaicumpa, W. Evolution of Therapeutic Antibodies, Influenza Virus Biology, Influenza, and Influenza Immunotherapy. *BioMed Res. Int.* **2018**, *2018*, 9747549. [CrossRef] [PubMed]
25. Dong, S.; Bo, Z. Screening for single-chain variable fragment antibodies against multiple Cry1 toxins from an immunized mouse phage display antibody library. *Appl. Microbiol. Biotechnol.* **2018**, *102*, 3363–3374. [CrossRef] [PubMed]

www.ingramcontent.com/pod-product-compliance
Lightning Source LLC
LaVergne TN
LVHW070611100526
838202LV00012B/623

MDPI
St. Alban-Anlage 66
4052 Basel
Switzerland
Tel. +41 61 683 77 34
Fax +41 61 302 89 18
www.mdpi.com

Foods Editorial Office
E-mail: foods@mdpi.com
www.mdpi.com/journal/foods